CAMBRIDGE MONOGRAPHS IN
EXPERIMENTAL BIOLOGY
No. 18

EDITORS:
P. W. BRIAN, G. M. HUGHES
GEORGE SALT (*General Editor*)
E. N. WILLMER

CIRCULATION IN FISHES

THE SERIES

CIRCULATION IN FISHES

BY

G. H. SATCHELL, B.Sc., Ph.D.

Professor of Biology, University of Sydney

CAMBRIDGE
AT THE UNIVERSITY PRESS
1971

CAMBRIDGE UNIVERSITY PRESS
Cambridge, New York, Melbourne, Madrid, Cape Town,
Singapore, São Paulo, Delhi, Tokyo, Mexico City

Cambridge University Press
The Edinburgh Building, Cambridge CB2 8RU, UK

Published in the United States of America by Cambridge University Press, New York

www.cambridge.org
Information on this title: www.cambridge.org/9780521279376

First published 1971
First paperback edition 2011

A catalogue record for this publication is available from the British Library

ISBN 978-0-521-07973-0 Hardback
ISBN 978-0-521-27937-6 Paperback

CONTENTS

vii

TO MY FIRST SON PAUL

PREFACE

This monograph has an explicit purpose. There are now several thousand papers dealing with the structure and function of the heart and blood vessels of fishes. It seems timely to attempt to integrate some of this information into a coherent account of the fish cardiovascular system. Only by assembling a jigsaw puzzle can we identify which pieces are missing.

Life began in water, most probably in sea water. The fish are the most numerous of all vertebrates. Their 20,000 or so species outnumber the summed totals of all the other vertebrate classes taken together. The aquatic environment covers more than 70% of the earth's surface and fish have lived within it for more than 350 million years. Any account of fish circulatory physiology is at present bound to do scant justice to the enormous variety of organisation and way of life present within the class Pisces. This must be so if only because of the uneven and fragmentary nature of the information available. The subject matter of the book is thus of necessity not evenly balanced.

Nevertheless, the book should prove useful to all who have an elementary knowledge of the mammalian circulatory system. It is aimed at helping biology students in their last year at high school and first two years at the university. The book will also, it is hoped, prove of interest to mammalian cardiovascular physiologists. Fish are red blooded aquatic gill-breathing vertebrates and their circulatory physiology demonstrates both primitive and specialised features which can illuminate our understanding of the better known mammalian condition.

I would like to accord my special thanks to the Senate of the University of Sydney for giving me time out to write. The initial development of the book occurred whilst I was Visiting Professor of the Zoology Department of the University of Bristol and I would like to thank Professor G. M. Hughes and his staff for their kindness and hospitality. I am grateful to Professor K. Johansen for his never failing help in discussing the text, and for the hospitality of his laboratory at the University of Washington. Dr D. J. Randall has many times come to my rescue with his unparalleled knowledge of the literature; I have been helped on many specific points by Dr D. Jones, Dr C. Lenfant, Dr D. Hanson and Dr G. Grigg. To all these gentlemen I accord my sincere thanks. I have made every attempt to acknowledge the work of others and I wish to apologise to any person whose work has inadvertently not been acknowledged. I would like to thank Mrs B. Jones and Miss M. Record for assistance with the typing and illustrations.

G. H. SATCHELL

August 1969

CHAPTER I

The Heart

Fish are aquatic vertebrates and most fish obtain their oxygen from the water in which they live. Whereas small invertebrate animals can rely solely on the physical process of diffusion to supply them with sufficient oxygen, in fish this is no longer possible as their larger size results in a lower surface:volume ratio. Diffusion is supplemented by a different physical process in which oxygen is transported by the mechanical propulsion of fluids, water or blood, which carry it along with them. This is the process of convection and we can recognise two convective systems (fig. 1). In one the surrounding water

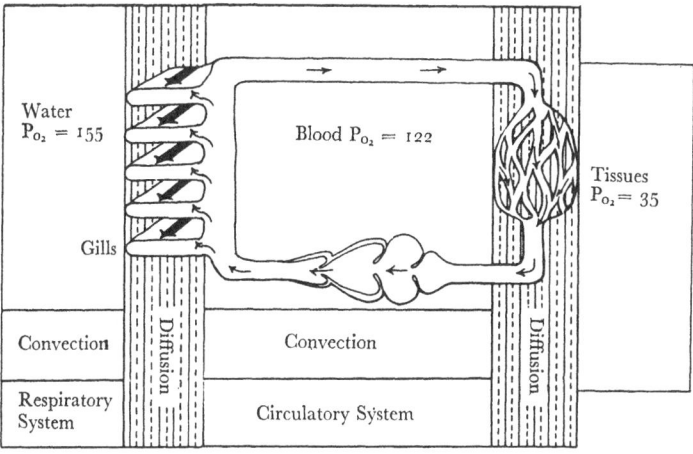

Fig. 1. The sites of convective and diffusional transport of oxygen in a fish. The partial pressures of oxygen quoted are those of aerated water, arterial and venous blood in *Salmo gairdneri*. Lower pressures probably occur within cells and within cellular organelles. Drawn partially after Holmgren (1966).

is pulled into the pharynx and forced between thousands of delicate blood-filled secondary lamellae of the gills and out through the expiratory openings. This is the respiratory system and is powered by the skeletal muscles of respiration. In the other, blood is forced, in turn through the secondary lamellae, distributing arteries, peripheral capillaries, and veins. It is powered by the cardiac muscle of the heart. Diffusion, however, is still the process which transports oxygen across the epithelia of the gill lamellae, and from the blood in the capillaries, through the tissue fluids into the mitochondria of the cells. The diffusional process is a passive one, and requires no energy other than that supplied by the convective systems which maintain an adequate pressure head. Despite the elaboration caused by the development of the two convective systems, we can still recognise the 'oxygen conduction line' (Holmgren, 1966) as involving a descending pressure gradient of oxygen from the environmental water to the mitochondria in the cells. The circulatory system constitutes the second of these convective systems and will be considered in the ensuing pages. It is, however, so intimately associated with the first of these, that we will have from time to time, to consider the respiratory system as well.

The position of the heart

Water contains relatively little oxygen compared with air. A litre of air at 15° C contains 209 ml of oxygen and weighs approximately 1·2 g. A litre of water contains 7·2 ml of oxygen and weighs 1·0 kg. It is clearly wasteful for a fish to transport water a greater distance than is absolutely necessary, and the gills are located at the rostal end of the body within the pharynx. The heart of fish is a gill heart. Apart from a few air-breathing fish, all of its output passes to the gills. It pumps only venous blood. The heart thus comes to be closely associated with the gills and is located immediately behind or between the posterior gill arches. It thus lies farther forward than in any other vertebrate. Cyclostomes are an interesting exception to this generalisation; their gill pouches are located more posteriorly than in the true fishes, and the heart is correspondingly farther down the body.

The pericardium

The heart is contained within the pericardium, cut off from the perivisceral coelom during development. Like it, the

2

pericardial coelom has two layers; the parietal pericardium forms the outer wall of the pericardial cavity, and the endocardium invests the heart. The parietal pericardium becomes adherent to adjacent cartilages and muscles which impart some rigidity to it. The semirigid pericardium imposes certain restraints on the contraction of the chambers of the heart and this affects cardiac function. It is therefore necessary that we should pay some attention to the pericardium before considering the physiology of the fish heart.

In elasmobranch fish the pericardial cavity is conical with its base formed by the fibrous pericardio-peritoneal septum. The ventral and lateral surfaces of the posterior part of the cavity are formed by the cartilage of the pectoral girdle. The roof is filled in with a triangular cartilage, the basibranchial plate of the pharyngeal skeleton. The conical rostral end of the cavity is supported by the powerful coraco-branchial muscles. The third, fourth and fifth pairs of these run parallel with and closely invest the cavity as they extend forward from the pectoral girdle to their insertions on the branchial arches. Depressor muscles of the branchial arches also share in forming the floor of the rostral part of the cavity. The pericardial cavity of skates and rays is flattened and is less rigid than that of the sharks and dogfish. In both it is true that the support imparted by the adjacent structures converts the pericardial cavity into a semirigid chamber that does not collapse when the heart is removed from it.

In teleost fish the rigidity of the pericardium varies greatly. In many, e.g. *Clarias lazera* (Nawar, 1955), the pericardio-peritoneal septum is moderately fibrous and is attached dorsally to the spinal column and ventrally to the coracoids. The lateral walls are fibrous partitions extending from the cleithra and coracoids to the fifth gill arch. Cleithro-pharyngeal and depressor-pharyngeal muscles complete the posterior walls of the cavity.

In elasmobranch fish, and in certain primitive bony fish such as *Acipenser*, the pericardial cavity communicates with the visceral cavity by a pericardio-peritoneal canal which arises dorsally above the atrium and runs through the septum along the ventral surface of the oesophagus. It is sometimes Y-shaped, with two openings into the visceral cavity. In the cyclostome *Petromyzon* the two cavities communicate during

3

the larval stage but are separate in the adult; in *Myxine* they remain in communication throughout life.

In the pericardial cavity is a small quantity of pericardial lymph which fills up the space between the heart and the pericardial wall and lubricates the movements of the atrium on the ventricle. In *Raja* (Smith, 1929) and *Squalus* (White and Satchell, 1969), the pericardial fluid is virtually free of protein. Smith believes it to be produced by the pericardial wall, presumably as an ultrafiltrate of plasma. The delicate walls of the pericardio-peritoneal canal collapse so readily that they act like a valve; coloured fluid injected into the pericardial cavity can be seen to pass easily through the canal and into the abdominal cavity, but fluids cannot be aspirated in the reverse direction.

Teleost fish lack the pericardio-peritoneal canal. In some genera, e.g. *Anguilla, Muraena, Cobitis,* the ventricle and to a lesser extent the atrium, are united to the dorsal wall of the pericardium by fibrous adhesions (Grant and Regnier, 1926). In some genera, e.g. *Protopterus,* the adhesions coalesce to form a single cord, the gubernaculum cordis. The conical ventricle of the blue marlin, *Makaira ampla,* is tethered to the posterior margin of the pericardium by three such cords (White, 1969).

The structure of the heart

The heart of fish consists of four chambers; the sinus venosus, atrium and ventricle are common to all fish (fig. 2). In lungfish both the atrium and ventricle are partly divided (Bugge, 1960). The fourth chamber, between the ventricle and the ventral aorta is represented by two rather different structures in the elasmobranch and teleost fish. Elasmobranch fish possess a conus arteriosus, a barrel-shaped chamber invested with cardiac muscle which contracts sequentially with the rest of the heart. Teleost fish possess a bulbus arteriosus, an onion-shaped elastic reservoir which is passively dilated by the blood ejected from the ventricle.

A   *The sinus venosus*

In most lower vertebrates the sinus venosus is the chamber in which the continuous inflow of blood from the great veins is first segmented into a pulsatile flow by the operation of the rhythmic activity of the heart muscle, and the sino-atrial

4

valves. In fish, the sinus lies dorsally across the back of the pericardial cavity and receives the ductus cuvieri of each side. In its posterior wall are the large openings of the hepatic veins, guarded in elasmobranch fish by sphincters of smooth muscle (Johansen and Hanson, 1967).

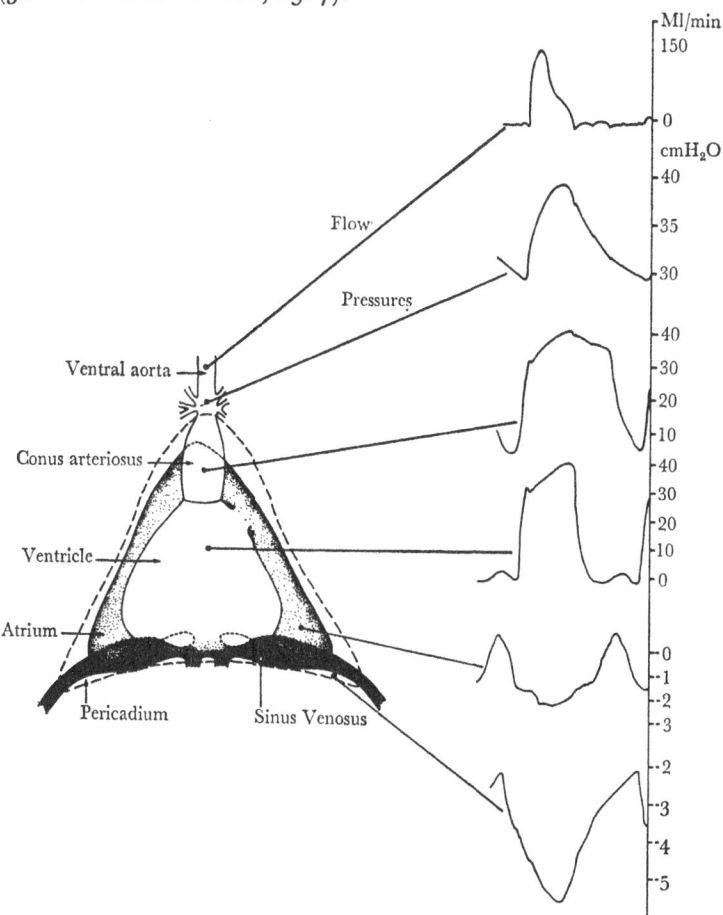

Fig. 2. The heart and pericardium of *Heterodontus portusjacksoni*. Pressure profiles from the pericardium and cardiac chambers, and the flow profile from the ventral aorta are presented to the right. The pressures were not recorded simultaneously.

The sinus is a thin-walled chamber with a delicate lining of cardiac muscle; its volume is never more than a fraction of that of the atrium into which it opens, although it would appear to be proportionately bigger in the hearts of certain Indian

cyprinid torrent fish such as *Orienus* (Saxena and Bakhshi, 1965).
The sino-atrial ostium is guarded by the large sino-atrial valve;
the two flaps are formed by a continuation of the smooth free
edge of the sinus wall into the atrium.

B *The atrium*
The single atrium is a large chamber lying above the ventricle
in elasmobranch fish, and rather more rostrally above the
bulbus arteriosus in certain teleost fish. In the Indian torrent
fish, *Botia birdi* (Cobitidae), the ventral surface of the atrium is
grooved to receive the bulbus arteriosus (Saxena and Bakhshi,
1965). The main dorsal chamber of the atrium is extended into
lobes which penetrate ventrally around the sides of the ventricle.
The atrium narrows to form a funnel-like passage which enters
the ventricle close to its junction with the conus. In some teleost
fish, e.g. Salmonidae, the atrio-ventricular ostium and the
ventriculo-bulbar ostium lie side by side separated by a narrow
band of tissue. The ventricular inflow and outflow tracts are
thus close together, and the ventricle is conical with its tip facing
posteriorly. Inside the atrium the delicate wall exhibits two
radiating fans of cardiac muscle fibres, the musculi pectinati.
Each arises from the atrio-ventricular ostium and arches
upwards, spreading out over the roof of the atrium and inter-
secting its fellow to form an acute-angled mesh. When these
contract they pull the atrial roof down towards the atrio-
ventricular ostium and empty the blood into the ventricle
(Benninghoff, 1933). Circular bands of cardiac muscle surround
the atrio-ventricular ostium and form a sphincter which supports
the atrio-ventricular valve. This consists of two membranous
flaps projecting into the ventricle and anchored to its walls by
fine threads. In *Protopterus* this valve consists of a plug of collagenous connective tissue with a nucleus of hyaline cartilage
(Bugge, 1960).

C *The ventricle*
The ventricle in elasmobranch fish is a thick-walled pyramidal
structure with its circular base resting against the pericardio-
peritoneal septum. In teleost fish it is usually conical with its
pointed tip facing posteriorly.
    There are two layers of ventricular muscle: the inner layer
consists of a spongy mesh of fibres projecting into the cavity and

the cortical layer forms the smooth outer surface of the ventricle. In eels and salmonid fish the spongy and cortical layers are separated by a layer of connective tissue. The coronary arteries are distributed only to the cortex. These two layers have not yet been characterised in any functional way. In *Gadus*, the cortical layer is missing. The cavity of the ventricle may be partially subdivided by the protruding trabeculae of the spongy layer; the cavity leads into the outflow tract directed rostrally towards the ventral aorta.

D   *The conus and the bulbus arteriosus*

The conus arteriosus is best developed in those fish in which the pericardium is most rigid. It is present in the Elasmobranchii, Holocephalii and certain primitive bony fish such as *Lepisosteus*, *Polypterus* and *Amia*. It is also present in the Dipnoi. It is surrounded by a layer of cardiac muscle, and internally bears three or four longitudinal ridges on which are borne one or more tiers of valves. Only the most distal row of valves are capable of spanning the lumen of the relaxed conus. Stohr (1876) recognised two types of valves, namely, typical pocket valves, and what he termed, tongue valves, which are protuberances lacking a cavity. *Cetorhinus* has four tiers of four valves, *Raja batis* has five tiers of three valves. *Lepisosteus* has the greatest number of any genus with 72 valves arranged in eight tiers. In some genera e.g. *Pristiurus*, the cardiac muscle only extends along the proximal half of the conus (Parsons, 1930).

The bulbus arteriosus is not invested with cardiac muscle; its wall consists of layers of smooth muscle and elastic tissue. It does not contract actively in sequence with the heart beat. However, the radiographic analysis of the eel heart by Mott (1950) shows that it can undergo periodic contractions typical of vascular smooth muscle. Some primitive families of teleost fish, e.g. the Albulidae, retain elements of both the conus and the bulbus arteriosus.

Electrical phenomena of the fish heart

The fish heart is a myogenic heart formed of typical vertebrate cardiac muscle fibres. Jensen (1965) has reported in the hagfish that the diameter of the muscle fibres in the atrium is 6·1 μm and in the ventricle is 7·1 μm. Perhaps cardiac muscle fibres in fish are smaller than those of other cold blooded vertebrates, e.g. frog ventricle 10 μm, turtle atrium 30–80 μm (Hoffman and Cranefield, 1960).

In the skate, *Dasyatis akajei*, atrial strips exhibit a resting membrane potential (r.m.p.) of 66·5 mV at 20° C (Seyama and Irisawa, 1967). In goldfish the r.m.p. of ventricular muscle is 72·6 and of atrial muscle 69·2 mV (Kuriyama *et al.* 1960). Cardiac muscle action potentials in the skate have the same form as in other vertebrates. The atrial muscle action potential has an amplitude of 92±2·6 mV at 20° C; the rate of rise is 9·5±0·7 V/sec. This is very low compared with that of a mammal, e.g. 560 V/sec for the Purkinje fibres of a sheep (Brady and Woodbury, 1960). It is low even for a cold-blooded vertebrate; in the frog ventricle it is 30 V/sec (Wiedmann, 1955). The low rate may be due to the relatively high concentration of intracellular sodium ions. That of the atrium is 91·7±17·4 mM/l, that of the ventricle is 110±10·6 mM/l. Because approximately half of the osmotic pressure of skate body fluid is contributed by ions other than sodium, the amplitude of the action potential is very sensitive to the external concentration of this ion.

The location of the pacemakers and nodal tissue in the fish heart is at present obscure. Jensen (1965) reported that many fibres in the heart of a hagfish exhibit pacemaker potentials in which the membrane potential falls towards a firing level and initiates an action potential when this is achieved. Laurent (1962) reported that in teleost fish the regions of the sino-atrial and atrio-ventricular junctions do not possess typical nodal tissue, but that scattered islets of such cells occur throughout the myocardium. However, the electrocardiograms of both *Heterodontus* (fig. 3) and *Anguilla* (Oets, 1950) show a V wave which originates from the sinus venosus and precedes the other events of cardiac excitation. Rybak and Cortot (1956) report that in *Scyllium* the isolated sino-atrial valve region will continue to beat. At least in these fish the sino-atrial node must be the pacemaker. Either the islet of pacemaker tissue in the node which has the fastest natural rhythm must entrain the others, or there is some mutual interaction between them. Grodzinski (1954) reported that the heart of the glass eel will beat apparently normally without its sinus venosus. Jullien and Ripplinger (1957) report that in carp, bass and other teleost fish a ligature between the sinus and atrium stops the contractions of the sinus whilst the atrium and ventricle continue to beat. A ligature between the atrium and ventricle immobilises the ventricle, though it remains excitable. This would seem to suggest that at least in some teleost fish the heart is either normally driven from an islet of pacemaker cells in the atrium, or can be paced in this way in the absence of its usual drive from the sinus venosus.

The mechanisms responsible for the delay in the passage of the cardiac impulse at the junctions between the chambers are unknown. Benninghoff (1933) has suggested that the circuitous pathways of the atrial fibres around the atrio-ventricular osteum may provide a delay. Visual observation of the beating heart in elasmobranch and in teleost fish (Randall, 1968) shows that the apex of the ventricle contracts before the basal region surrounding the atrio-ventricular ostium. As the ventricle is activated in this region, fast-conducting intraventricular fibres must exist; the study by Chiesa, Noseda and Marchetti (1962) of the activation of the ventricle in *Salmo*, *Anguilla*, *Cyprinus* and *Ictalurus* shows that the left apical region is activated before the right.

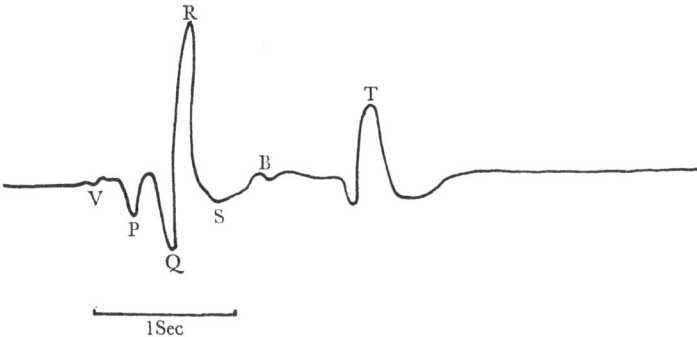

Fig. 3. The electrocardiogram of *Heterodontus portusjacksoni* (after Holst, 1969).

Electrocardiograms (ECG) of fish have now been known for more than fifty years (Zwaardemaker and Noyens, 1910). The amplitude of the ECG is usually small and leads are usually taken from the skin or tissues just below the heart. Kisch (1948) reported two additional events present in elasmobranch fish that do not occur in the ECG of mammals. These are the

V and B waves (fig. 3). The V wave is the initial deflection; it is caused by the depolarisation of the sinus venosus. It has been recorded in both *Heterodontus* and in the eel (Oets, 1950). The V wave also occurs in the ECG of the frog, where it arises from the sinus.

The P wave signals the depolarisation of the atrium. Labat (1966) reported a Ta wave that signals the repolarisation of the atrium in teleost fish; normally it is obscured by the large QRS complex. The PR interval represents the delay in the passage of the impulse from the atrium to the ventricle.

The QRS deflection signifies ventricular depolarisation. Q is always a negative deflection, R is positive; the S deflection is usually feebly developed. Oets (1950) reported that the QRS deflection of the eel is briefer than that of mammals. The QT interval during which the ventricular muscle remains depolarised, endures for longer than that of a mammal: Table 1 (Labat, 1966) gives the durations of these intervals for six species of teleost fish. The T wave is most often positive and has a rounded form; it signals the repolarisation of the ventricle and the relaxation of the ventricular muscle.

TABLE 1. *The time intervals of the electrocardiograms of six species of fish at 18° C, based upon 100 measurements (from Labat, 1966)*

	Catfish	Tench	Carp	Sole	Eel	Mullet
P	0·08 s ±0·01	0·08 s ±0·01	0·06 s ±0·01	0·06 s ±0·01	0·04 s ±0·01	0·04 s ±0·01
PR	0·24 s ±0·02	0·16 s ±0·01	0·14 s ±0·01	0·12 s ±0·01	0·08 s ±0·01	0·08 s ±0·01
QT	0·45 s ±0·02	0·46 s ±0·02	0·40 s ±0·02	0·24 s ±0·02	0·20 s ±0·02	0·20 s ±0·01

During the QT interval in elasmobranch fish there occurs a small signal called by Kisch (1948) the B deflection. It signals the depolarisation of the conus arteriosus. It has not been reported in teleost fish which lack this muscular chamber. Tebecis (1967) has made focal recordings along the length of the conus arteriosus of *Heterodontus* and shown that the B wave

10

moves towards the rostral end with a velocity of 2–4 cm/sec at 18° C. Sometimes it is possible to see a separate Br wave following the T wave; it signals the repolarisation of the conal muscle.

The various time intervals of the fish ECG are lengthened at lower temperatures. Nevertheless at the same temperature there exist specific differences (table 1). Active fish like mullet and eel have shorter ECG time intervals than slowly-moving species like catfish and carp.

The events of the cardiac cycle

There have been a number of studies of pressure and flow within the chambers of the fish heart during the last ten years. The hearts of elasmobranch fish have received most attention. Johansen (1965a, b) has recorded intrapericardial pressures in *Raja binoculata* and *Squalus suckleyi*; Sudak (1965a, b) has recorded intrapericardial and intracardiac pressures in *Mustelus canis*; Satchell and Jones (1967) have recorded pressures and flow within and out of the conus arteriosus of *Heterodontus portusjacksoni*. Teleost fish are less well known; Randall (1968) has published records obtained by Stevens and Bennion from the heart chambers of lingcod, *Ophiodon elongatus*. The hearts of the three genera of lungfish *Neoceratodus*, *Lepidosiren* and *Protopterus*, have been studied by Johansen, Lenfant and Hanson (1968) and Johansen and Hanson (1968) with special reference to the evolution of air-breathing in these fish. Pressure changes within the aneural heart of the cyclostome Myxine have been described by Johansen (1960).

A major difference between the heart of an elasmobranch fish and that of a terrestrial vertebrate is that the pressure within the pericardial cavity is below the ambient pressure in the tissues surrounding it. Schoenlein (1895) recorded intrapericardial pressures of −2 to −5 cm H_2O in *Torpedo ocellata* and suggested that the heart aspirated blood from the veins. Brunings (1899) emphasised that the elasmobranch heart operates both as a pressure pump and a suction pump. More recently Johansen (1965b), Sudak (1965b) and Satchell (1969) have validated the claims of the earlier workers. As the blood is ejected from the ventricle into the ventral aorta, the total volume of the contents of the pericardium decreases, and the semirigid walls are tensed inwards. This generates a negative intrapericardial pressure which can only return to its end diastolic level when the atrium has received an equivalent volume of blood. The negative pressure is thus due to a lag between the ejection of blood from, and return of blood to the pericardium. The slower the heart rate, the more completely does the intrapericardial pressure return towards zero pressure, i.e. towards the ambient pressure in the tissues surrounding the pericardial cavity. But at normal heart rates the return of blood to the heart in diastole is not sufficiently completed to enable the inwardly stressed parietal pericardium to expand back to its resting position before the next ventricular ejection has once more decreased the volume of the contents of the pericardial cavity and lowered the pressure again. Some aspects of this topic will be discussed again; it is necessary to introduce it at this stage because it explains why in elasmobranch and some teleost fish the diastolic pressure within the sinus venosus and atrium are below environmental pressure.

The contraction of the sinus venosus is signalled by the occurrence of the V wave of the ECG. Visual observation of the sinus within the opened pericardium demonstrates that it does indeed contract, but in elasmobranch fish the contraction is a feeble one, and no worker has yet succeeded in recognising any elevation of pressure from records obtained either from the sinus or from the ductus cuvieri. Such records demonstrate that the sinus normally sustains a pressure of -2 to -7 cm H_2O (Sudak, 1965a, b; Johansen and Martin, 1965), and is at a slightly higher pressure than the atrium. This is to be expected as the sinus is placed along a pressure gradient that has its lowest point within the atrium and extends uphill to zero pressure some distance along the great veins. In the lingcod (Randall, 1968) the pressure within the sinus venosus is again a little above that of the atrium but is positive with respect to the surrounding tissues. Here, moreover, there is a rise in pressure in the sinus venosus immediately preceding atrial systole, suggesting that the contraction of the sinus plays a greater role in atrial filling than in elasmobranch fish.

Atrial systole occurs following the P wave of the ECG. Atrial pressures in *Mustelus* (Sudak, 1965a), *Heterodontus* and the lingcod (Randall, 1968) are below ventricular pressures throughout the cardiac cycle except for a brief period towards the peak of atrial systole (fig. 2). At this time records from within the ventricle and the conus arteriosus demonstrate that pressures within them rise to levels just below those exhibited by the atrium. Visual inspection of the ventricle and conus arteriosus shows that they expand during atrial systole. In *Heterodontus* the flattened base of the pyramidal ventricle moves posteriorly as the blood is transferred from one chamber to the other. The intrapericardial pressure does not at this time exhibit any appreciable fall, suggesting that the sino-atrial valves close and are competent, so that no blood refluxes through them and leaves the pericardial cavity. If, however, the intrapericardial lymph is deliberately aspirated from the pericardial cavity, thereby causing the delicate atrium to stretch and fill the additional space, then atrial systole is accompanied by a fall in intrapericardial pressure (Holst, 1969). Presumably the sphincter surrounding the sino-atrial valve cannot ensure the competence of the flap valve when subject to this unusual stretch.

The QRS deflection of the ECG signals the onset of ventricular systole, and the rise in pressure in the ventricle closes the two pocket valves that guard the atrio-ventricular opening. The event is sometimes signalled by a distinct valvular oscillation on the ventricular pressure trace (Satchell and Jones, 1967).

Neither the elasmobranch nor the teleost heart shows a rise in atrial pressure during the onset of ventricular systole such as occurs in mammals where it is attributed to the bulging of the atrio-ventricular valves. The contraction of the muscular sphincter in the fish heart which surrounds these valves, we may suppose, reduces the area they have to span and prevents this. In the mammalian heart the openings of the atria into the ventricles are large and surrounded by fibrous rings; the valves are consequently large and present a greater proportionate surface area facing into the atrium.

As tension increases in the ventricular muscle some of the blood is caused to flow into the conus arteriosus. Of the three sets of valves in this structure, it is the upper tier that is closed during the latter half of diastole, and its musculature has not at this stage been activated. Consequently, it undergoes a further passive dilation (Satchell and Jones, 1967; Sudak, 1965a; March, Ross and Lower, 1962). In teleost fish the single tier of valves is located at the bottom of the elastic bulbus arteriosus, so that no blood can pass from the ventricle to the bulbus until these open. When ventricular pressure exceeds that in the ventral aorta, the bulbal valves open and ventricular ejection commences.

It is clear that the elasmobranch ventricle does not strictly have a period of isovolumic contraction, as does that of a mammal. There is no period between the closure of the atrio-ventricular valves and the opening of the upper valve in the conus arteriosus during which the volume of the ventricle remains the same whilst the pressure within it rises. Teleost fish, in contrast, do exhibit a period of isovolumic contraction as the blood in the ventricle cannot pass the tier of valves in the bulbus arteriosus until ventricular pressure exceeds ventral aortic pressure. When the valves open, the period of rapid ejection commences. Records of the velocity of blood flow in the ventral aorta demonstrate, in *Squalus* and *Raja* (Hanson, 1967), *Heterodontus* (Satchell and Jones, 1967), and in the lingcod (Randall, 1968), a sudden rapid increase in the velocity of flow. In *Heterodontus* this is preceded by a very small increase that perhaps reflects the forward movement of the valve ring during ventricular systole. Following the initial increase the velocity falls again and in *Heterodontus* returns to the zero level as the lower tier of valves close. Ventricular pressure rises to a peak towards the end of ventricular systole (fig. 2) indicating that the compliant reservoir of the ventral aorta and afferent branchial arteries have undergone a progressive distention, as blood poured into them in excess of outflow.

The expulsion of blood from the ventricle into the ventral aorta diminishes the volume of the heart and causes a drop in pressure within the pericardium (fig. 2), and therefore within the atrium and sinus venosus. This diminution in the volume is, we may suppose, in part offset by an accelerated inflow of blood from the veins. However, the pericardial pressure falls throughout the period of ventricular ejection; this corresponds with the time during which the pericardio-peritoneal septum can be seen to be tensed inwards into the pericardial cavity.

The B wave of the ECG indicates the onset of systole in the conus arteriosus and the base of the chamber constricts. This occurs before ventricular systole has quite finished, and serves to close the lower set of valves. It is presumed that this occurs in part through the apposition of the cusps of the lower tier of valves, and in part by the backflow generated within the chamber from the constriction of the stream of blood above these valves. The closure of the lower valves prior to the decline of pressure in the ventricle is known because it causes a characteristic valvular oscillation on the pressure records taken in the conus and ventral aorta (Sudak, 1965a; Satchell and Jones, 1967). March *et al.* (1962) have also obtained evidence of this in *Triakis* by high speed cinematographic analysis.

Following the closure of the lower tier of valves the wave of contraction passes up the conus arteriosus and causes the closure of the middle tier of valves (fig. 2) again just prior to the opening of the lower ones. Finally, the contraction dies away, the middle valves open as the upper ones close, and the pressure in the upper part of the chamber falls away to that prevailing in the ventricle. This is often below environmental pressure since the ventricle is subjected to the low pressure within the pericardial cavity.

The pressure trace in the ventral aorta exhibits at least two oscillations. The first of these is synchronous with the closure of the lower tier of valves, the second signals the closure of the upper valves. Sometimes there is a small oscillation between these that occurs as the middle tier of valves closes.

The record of the velocity of ejection in the ventral aorta demonstrates that in *Heterodontus* most of the outflow is confined to the period between the opening of the upper valves, and the closure of the lower valves. There is little if any backflow associated with this closure (fig. 2). Pressure records taken between the lower and middle, and the middle and upper valves of the conus arteriosus support this interpretation. Following the closure of the lower valves the whole chamber is in communication with the ventral aorta and shares its gradual decline of pressure as blood passes through the gills and into the peripheral circulation. The pressure in the lower chamber thus remains elevated for a period after that in the ventricle has fallen away.

The closure of the lower and middle valves is dependent upon the contraction of the cardiac muscle fibres that encircle the conus. The delay in the passage of the wave of electrical activity from the apex of the ventricle to the base of the conus arteriosus is such that the lower tier of valves is apposed before any backflow caused by the decline of ventricular pressure has occurred.

15

If the local anaesthetic MS 222 is placed on this junction the passage of the electrical activity is delayed and the fall of ventricular pressure occurs just before valve closure, permitting a brief interval of backflow from the ventral aorta into the ventricle (Satchell and Jones, 1967). More complete anaesthetisation of the conus arteriosus and its muscle may temporarily prevent any of the valves, except the upper ones, closing. These are large enough to close by backflow even in diastole. Indeed they are perfectly competent in the dead animal. But closure in the absence of any supporting systole of the conus arteriosus is accompanied by a considerable backflow from the ventral aorta into the ventricle.

The effects of pericardial pressure on the events of the cardiac cycle

A *Atrial filling*

The mechanism of atrial filling in elasmobranch fish contrasts with that of mammals where the *vis a tergo* of the blood in the venae cavae and pulmonary veins is sufficient to extend the atrial muscle during atrial diastole. The fact that in *Heterodontus* atrial pressure is below environmental pressure throughout diastole suggests that this sucking pressure generated within the pericardium aspirates blood into the atrium. The gradient of pressure between the peripheral veins and the atrium is due, it would seem, to a combination of *vis a tergo* and *vis a fronte*. Hanson (1967) has shown that opening the pericardium of *Hydrolagus* depresses both cardiac output and ventral aortic blood pressure. In the absence of cardiac suction venous pressure must rise above environmental pressure if the atrium is to be filled by passive inflow.

The posterior cardinal sinus, it has been suggested (Holst, 1969), has a specific function in providing a reservoir immediately behind the heart from which blood can be aspirated as the atrial pressure falls during ventricular systole. In its absence, energy would need to be expended in accelerating blood along narrow tubular channels. The posterior cardinal sinus, it seems, enables much of the return to the heart to be compressed into the period of ventricular systole. The well developed sino-atrial valve is probably also a necessary feature of a system of atrial filling dependent in part upon pericardial suction. During atrial systole there is a residue of negative

16

pressure remaining in the great veins and in the absence of such a valve, the blood would be sucked back into the posterior cardinal sinus.

B *Ventricular filling*

The muscular walls of the cardiac chambers exert a slight tension even when relaxed and it is to be expected therefore that during diastole the pressure within the thickly muscled ventricle will be higher than that in the surrounding pericardium. The records of Sudak (1965a) for *Mustelus* show this quite clearly. Only during atrial systole does atrial pressure exceed ventricular pressure, and the atrio-ventricular valves are probably closed during the remainder of the cardiac cycle. The atrium must therefore be largely if not solely responsible for ventricular filling in elasmobranch fish, and Randall (1968) notes that the atrial volume in both elasmobranch and teleost fish is approximately equal to that of the ventricle. We can contrast this with the condition in mammals. Mitchell *et al.* (1962) report that atrial systole increased ventricular output only 50% over that maintained by venous pressure alone. The *vis a tergo* of the blood in the venae cavae serves to move blood through the atrium into the ventricle during diastole. In mammals atrial systole is never more than an auxiliary means of ventricular filling.

In elasmobranch fish the atrium and ventricle are thus involved each in assisting the filling of the other in a reciprocal manner. Harvey specifically commented on this aspect of the fish heart in his second letter to John Riorlan (1649) and likened the reciprocal action of the atrium and ventricle to the operation of antagonist muscles at a joint. The measure of reciprocity that occurs in fish is indeed due to the fact that the two chambers are enclosed within the semirigid pericardium.

The function of the conus arteriosus

The conus arteriosus of elasmobranch fish serves, it has been suggested, as an additional pump that extends the flow of blood to the ventral aorta and gills into the latter part of the cardiac cycle (March *et al.*, 1962). The flow of blood within the ventral aorta has been recorded in *Squalus suckleyi*, *Raja binoculata*, and *Heterodontus francisci* (Hanson, 1967; Johansen,

Franklin and Van Citters, 1966), and such records do show a small acceleration of the blood following ventricular systole, attributable to the conus. Satchell and Jones (1967) suggest that the conus serves to postpone the closure of the upper valve by backflow from the ventral aorta until the negative pressure within the pericardial cavity has abated by atrial filling. It was argued that a negative pressure applied outside a valved chamber might lead to a backflow which would impair the efficiency of the counter-current exchange of oxygen between the blood and the respiratory water in the gill lamellae. Randall (1968) remarks that both in elasmobranch and teleost fish flowmeter records show that the aortic valves are closed without backflow, unlike the aortic valves of mammals. Without the delay imposed by the conus arteriosus, the upper tier of valves would be closed by backflow immediately following ventricular systole, at a time when the intrapericardial negative pressure was greatest. In *Heterodontus*, Satchell and Jones (1967) found that paralysing the conal musculature with local anaesthetic caused this to happen and increased the backflow in the ventral aorta from $6.8 \pm 2.2\%$ to $30.8 \pm 5.4\%$ of forward flow.

Perhaps the conus arteriosus serves both functions. The records of *Heterodontus* were made on restrained inactive fish; it is a bottom-living species and the pumping function of the conus may be less important.

The function of the bulbus arteriosus

In elasmobranch fish the conus arteriosus terminates just before the apex of the pericardial cavity so that a short length of ventral aorta remains within it. In teleost fish it is this intrapericardial portion of the ventral aorta that balloons out to form the thick-walled elastic onion-shaped chamber, the bulbus arteriosus. The conus arteriosus intervening between it and the ventricle, is reduced. Some primitive families of teleost fish such as the Albulidae retain a small conus arteriosus with two tiers of two valves; in some others, e.g. *Dorosoma*, it is reduced to a single tier of two or three valves surrounded by a ring of cardiac muscle (Smith, 1918).

The bulbus arteriosus is located beneath the atrium. It is rich in elastic tissue; Lansing (1959) remarks of the bulbus arteriosus of *Lophius piscatorius*: 'The organ is highly elastic, more so than any biological material previously encountered.' Radiographic analysis of the eel heart (Mott, 1950) shows that the contraction of the ventricle inflates the bulbus arteriosus. It increases in diameter throughout ventricular systole. Johansen (1962) has shown in *Gadus* that the bulbus arteriosus acts as an elastic reservoir, i.e. a 'windkessel'. The rapid rise and fall of pressure within the ventricle is damped by the bulbus so that blood pressure recorded from within it does not achieve its peak until that of the ventricle has started to fall. The velocity of flow in the ventral aorta is also affected by this elastic reservoir; in both the cod, *Gadus morrhua* (Johansen, 1962) and the lingcod, *Ophiodon elongatus* (Randall, 1968), the initial acceleration following ventricular systole is not so rapid as in an elasmobranch fish, and the flow continues at a reduced level almost to the end of diastole. Zero flow exists in *Gadus* only for a brief period. Thus the bulbus arteriosus though not a pumping chamber, serves, like the conus arteriosus of elasmobranch fish, to extend the proportion of the cardiac cycle during which blood flows into the gills.

The radiological analysis of the eel heart by Mott (1950) suggests that the incorporation of the elastic bulbus arteriosus into the pericardium may serve another function. It is, in some genera, e.g. *Salmo*, placed below the atrium which may be grooved on its ventral surface so that the two fit precisely together. Mott (1950) reports that the emptying of the bulbus is accompanied by the filling of the atrium. The two structures fill and empty reciprocally. Mott has shown (1951) that pressures of -5.4 to 6.7 cm H_2O can be recorded in the hepatic vein of the eel, suggesting the cardiac suction occurs in some teleost fish. The elastic reservoir of the bulbus arteriosus permits the outflow of blood from the heart to be spread more evenly through the cardiac cycle, and thus the filling of the atrium to be correspondingly extended. Perhaps this helps to explain why the posterior cardinal vein in teleost fish is a vein with a circular cross section and not an expansile venous sinus. If the rapid development of pericardial negativity that characterises the ejection from the elasmobranch ventricle is in teleost fish spread more evenly through the cardiac cycle, there would be no need for a reservoir of blood to permit a rapid atrial filling (Holst, 1969).

CHAPTER 2

The Arteries

The studies of Hughes (1960), Hughes and Shelton (1958, 1962), Hughes and Ballintijn (1965, 1968) and Hughes and Umezawa (1968) have provided us with an account of the operation of the respiratory pumps in a number of elasmobranch and teleost fish. The stream of water that flows between the secondary lamellae of the gills is propelled by a double pump. The buccal or orobranchial component is a pressure pump that forces water from the pharynx between the secondary lamellae. The opercular or parabranchial component is a suction pump that aspirates water into the parabranchial cavity from the gill lamellae. The two pumps act in sequence and maintain a flow of water throughout a greater proportion of the respiratory cycle than either alone could achieve. The terms inspiration and expiration are not really applicable to fish; they refer to the tidal movement of the respiratory medium in terrestrial vertebrates. In fish the respiratory medium is propelled continuously in one direction.

A heart powered by cardiac muscle and equipped with valves must inevitably produce a pulsatile outflow. It might thus be expected that the arterial system would be so modified as to smooth this and match the even flow of water past the secondary lamellae by an even flow of blood through them.

Taylor (1964) has discussed the design criteria of a model circulatory system and depicts the mammalian circulation by a 'one windkessel model' (fig. 4a). The capacitance of the arterial system CP representing the compliance of the aorta and arteries, is charged up rapidly during systole and discharges during diastole, through the peripheral resistance RP.

In fish, the capacitance CP (fig. 4b) must receive its charge through the resistance of the gills RG; a sudden surge early in each cycle would be inevitable if the chief capacitance were located beyond the gills. To diminish this surge, in the interests of achieving a more even flow through RG, it is necessary to diminish the capacitance of CP as much as possible, and place a new smoothing capacitance, CG, between the heart and the gills. It must be big enough to achieve an acceptable smoothing of flow through the gills, and this will demand that it be large compared with CP. This capacitance could be provided by the distensibility of the ventral aorta and afferent branchial arteries. It is desirable that the time constant of the RC combination comprising the capacitance of CG and the impedance of the rest of the circuit, be large compared with the period of the cardiac cycle. However, it must not be excessively

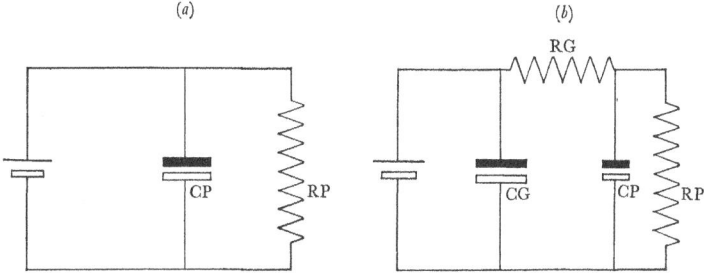

Fig. 4. Models of (a) the mammalian, and (b) the fish circulatory system. RP = resistance of peripheral vessels. RG = resistance of gill vessels. CP = capacitance of peripheral vessels. CG = capacitance of ventral aorta and afferent branchial arteries.

large, or the heart will have difficulty in changing the blood pressure, as it will have to pump a large quantity of blood from the venous reservoirs into the compliant reservoir ahead. The value of CG must be a compromise between these two opposing demands. The fact that the dorsal aortic blood flow is pulsatile shows that smoothing is not complete (Randall, Jones and Shelton, 1969).

Theoretical deduction would thus suggest that the model fish circulatory system will have a large and compliant ventral aorta, and a rigid nondistensible dorsal aorta. The heart should also beat at such a rate that diastole occupies only a brief interval in each cardiac cycle. The reduction of diastole will diminish the pulse pressure and the pulsatile nature of the inflow into the arteries. However, as Taylor (1964) has pointed out for the mammalian heart, there is a strict limit imposed on the reduction of diastole, by the necessity of achieving an adequate flow through the coronary arteries. Coronary flow is effectively limited to diastole because the tension within the myocardium during systole closes the coronary vessels. This restriction we may suppose is greater in the fish heart because it pumps only venous blood, and cannot perfuse its coronary arteries with the full force of the ventricle; it obtains its coronary supply from the efferent branchial arteries where the blood is oxygenated but the pressure has been diminished by the resistance of the gills. We may therefore expect that it is in the histology of the arterial wall, and in particular in the proportion of the two important structural proteins collagen and elastin, that the changes which will smooth the branchial flow are to be sought.

The proportion of elastin and collagen in the arterial wall

Mammalian studies show that the elastic modulus of the arterial wall is closely related to the proportions of elastin and collagen in it. Although these structural proteins can be recognised histologically by classical staining techniques they can only be identified quantitatively by chemical analysis. The determination of elastin and collagen depends on the fact that collagen can be gelatinised in water in an autoclave and thereby separated from elastin. The collagen can be estimated as the oxidation product of the hydroxyproline residues that characterise this protein; the residue remaining in the autoclave, when freed from other tissue constituents with sodium hydroxide, is elastin (Neuman and Logan, 1950).

Lander (1964) has analysed the arterial walls from two grey nurse sharks, *Odontaspis arenarius*. He removed four samples from the base to the bifurcation of the ventral aorta, one from each of the afferent and efferent branchial arteries and three along the length of the dorsal aorta. He performed similar analyses on single specimens of a bronze whaler shark, *Carcharinus ahenea*, a hammer-head shark, *Sphyrna lewini*, and a white pointer shark, *Carcharodon carcharias*. The data from one of the grey nurse sharks are presented in fig. 5. They agree with those of the other three species in showing a gradient of diminishing elastin content from the proximal end of the ventral aorta through the afferent and efferent arteries up to the dorsal aorta.

In the white pointer shark the fourth afferent branchial artery was subdivided into three portions and assessed separately. There was a gradient of elastin content along its length, the proximal, i.e. cardiac, end exhibiting 19·5% and the distal end 10·8%. The dorsal aorta showed no significant changes in elastin content along its length, but it was, in each species, richer in collagen than the ventral aorta; there was an increasing gradient of collagen content from the one to the other.

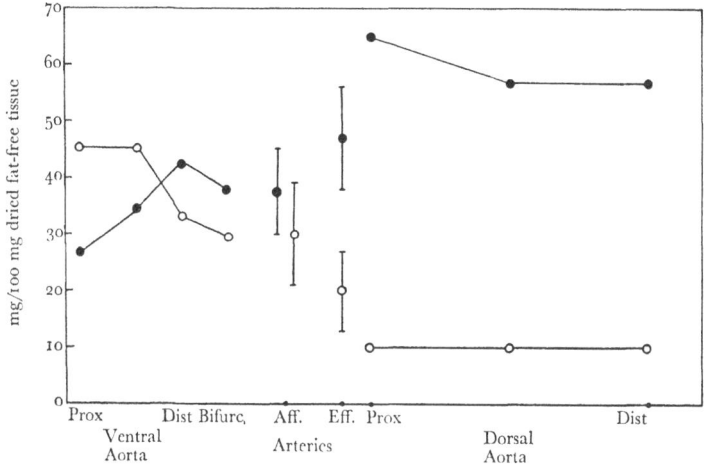

Fig. 5. The amounts of collagen and elastin expressed as mg/100 mg of dried fat-free tissue, in the main arteries of *Odontaspis arenarius*, the grey nurse shark. ● collagen. ○ elastin. After Lander, 1964.

The bronze whaler shark had the most collagenous dorsal aorta of the four species; collagen constituted 93·4% of the dried fat-free weight. Analyses along the length of the dorsal aorta showed no consistent pattern of change in collagen content in the four species. An average of the determinations of all five specimens together reveals that in the ventral aorta elastin constituted 31·0% and collagen 46%, whilst in the dorsal aorta elastin constituted 9% and collagen 69%. The ventral aorta has approximately three times as much elastin in its wall as the dorsal aorta. Nevertheless, collagen is quite abundant in the ventral aorta.

The sum of collagen plus elastin content is of interest. As mucopolysaccharides constitute only 2% of the arterial wall (Kirk, 1959) the remaining tissue constituent of the dried

23

fat-free wall must be smooth muscle. This bears out the histo-logical findings, that there is a variable amount of smooth muscle in the walls of the vessels, mostly greater than 20%, and that the afferent and efferent arteries are more muscular than the aortae. The differences were more marked in the data for the bronze whaler and white pointer sharks.

The histology of fish arteries

Lander (1964) has described the histology of the arteries that he analysed. The ventral aorta (fig. 6a) has the typical three layers of the vertebrate arterial wall but the media constitutes seven-eighths of the total. It consists of circularly arranged elastic fibres with smooth muscle cells in between. The intima which stains positively for collagen forms a delicate scalloped

(a)   (b)

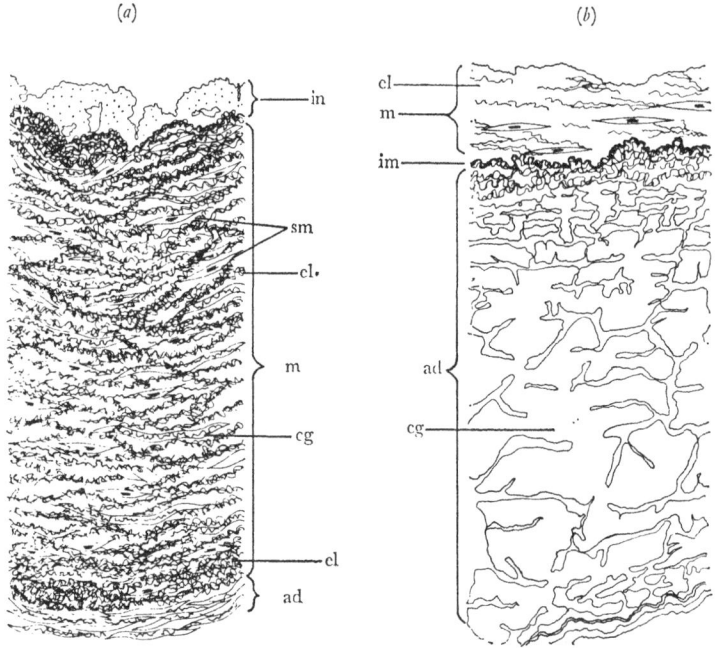

Fig. 6. Schematic diagrams of transverse sections of (a) the ventral aorta, (b) the dorsal aorta of *Odontaspis arenarius*. (ad = adventitia, cg = collagenous fibres, el = elastic fibres, in = intima, im = internal elastic lamina, m = media, sm = smooth muscle cells.) (After Lander, 1964.)

lining layer. The adventitia forms a thin investing outer layer containing collagen fibres.

The wall of the dorsal aorta (fig. 6b) has a bilamina structure; no separate intimal layer is present. The media is rich in fine elastic fibres; if stripped off from the rest of the wall and analysed separately it shows a higher proportion of elastin than the whole wall (10·5% rather than 5·7%). There is a fine internal elastic lamina. The media also contains some smooth muscle fibres. The adventitia constitutes three-quarters of the wall and is extremely collagenous with oblique fibres; some, nearer the media, run round the vessel.

The afferent and efferent arteries have structures intermediate between those of the ventral and dorsal aortae. Afferent vessels have an abundance of circularly arranged elastic fibres with some smooth muscle fibres in the media as well. Efferent vessels have fewer elastic fibres but are richer in smooth muscle. Dornesco and Santa (1963) have described the histology of the blood vessels of the carp; their account agrees in its essentials with that of Lander.

The elastic modulus of fish arteries

Learoyd (1963) has determined the distensibility curves of the ventral aorta, the afferent and efferent arteries and the dorsal aorta of two of the species of shark studied by Lander (1964), *Odontaspis arenarius* and *Carcharodon carcharias*. The curves for these four vessels are derived from the two species taken together (fig. 7).

Roach and Burton (1957) studied the contribution that elastin and collagen make to the elastic modulus of a vessel wall by digesting away each protein separately and determining the deformation of the wall in response to pressure. They showed that elastin has a much lower elastic modulus, 30 N/cm²/ 100% elongation, than collagen, 1×10^4 N/cm²/100% elongation, and that the initial response to deformation at relatively small distending pressures reflects the stretching of elastin. The steeper slope that follows this occurs as the collagen fibres are progressively extended. These findings suggest that the low value of the elastic modulus of the ventral aorta, and the gentle slope of its curve (fig. 7) reflect the high proportion of

25

elastin in it, and suggest that at the pressures it is likely to encounter, its elastic modulus is dominated by the elastic fibres within it, rather than the collagen it contains. The dorsal aorta, with its larger elastic modulus, reflects its richness in collagen; the increasing slope of the curve is characteristic of a vessel in which distention brings about a rapid increase in tension in the rather inextensible wall.

It is instructive to compare the dorsal aorta of a shark with that of a mammal. The shark aorta contains at least twice

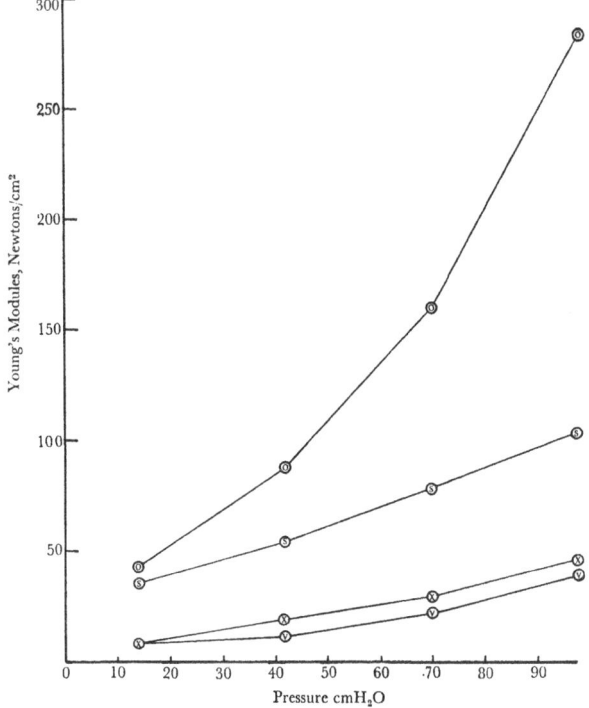

Fig. 7. The modulus of elasticity of shark arteries based upon combined data from *Odontaspis arenarius* and *Carcharodon carcharias*. (v = ventral aorta, x = afferent branchial artery, s = efferent branchial artery, o = dorsal aorta.) (After Learoyd (1963).)

as much collagen as the most collagenous part, i.e. the femoral end, of a mammalian aorta (Taylor, 1964). If we consider its elastic modulus at a particular pressure, e.g. 70 mmHg, it is $270 \, \text{N/cm}^2$, which is at least twice that of the femoral end of the mammalian aorta (Lander, 1964). Moreover, the dorsal aorta of a fish exhibits no obvious gradient of collagen content as does the aorta of a mammal. It is in no sense a 'tapered transmission line'. Lander (1964) reports that there is no amplification of the pulse wave along the length of the aorta such as occurs in mammals.

This rigid collagenous vessel can scarcely be regarded as an evolutionary response to the need to withstand high internal distending pressures. Rather does it seem, as suggested in the introduction to this chapter, that its rigid wall serves to minimise the windkessel properties that such a long and branched vessel must have, and thus minimise the initial rush of blood through the gills in the ejection phase of each cardiac cycle. Furthermore, these studies show that the ventral aorta has indeed the histological structure and elastic modulus that equip it to serve as a windkessel.

Structural differentiation in the dorsal aorta

The dorsal aorta runs below the spinal column, is attached firmly to it, and flexes with it as the fish swims. This close link between the two structures has brought about changes in the aorta so that it is no longer a uniform tubular vessel but is differentiated both along its length and around its circumference.

The aortic wall is much thinner where it abuts against the ventral surface of the vertebrae and only a few strands of the dense collagenous tissue of the adventitia are present there. Circular muscle fibres are absent in this region, sparsely present at the sides, and more abundant on the ventral surface of the wall. The adventitia on the ventral surface of the aorta is, in each segment, extended laterally to form a girdle of strands that lashes the aorta to the vertebrae. Dornesco and Santa (1963) report that this girdle is present in the middle of each vertebra and thins at the junction between them.

In some teleost fish the dorsal wall of the aorta is thickened to form a median ridge that projects into and partly divides the lumen of the vessel (fig. 8a, b). In the carp (Dornesco and Santa, 1963) it is most extensive below the middle of each vertebra, and is less evident at the junctions between them. It is present also in the shad, *Clupea alosa*, and Burne (1909) suggested that its sinuous movements propel blood along the aorta. This idea is supported by De Kock and Symmons (1959) who noted this dorsal ridge in herring and trout and reported that bending the body caused the median ridge to form a diagonal curtain passing down the length of the aorta. Favaro (1906) described this structure in *Acipenser* and *Tinca*, and termed it the hypochordal ligament. No functional studies of it have yet been made.

In carp (Dornesco and Santa, 1963), *Scyllium, Tinca, Atherina, Mugil* and *Ophichthys* (Favaro, 1906), *Labrus* and *Crenilabrus* (Laguesse, 1892), and in *Petromyzon* (Wagenvoort, 1952), the segmental arteries leaving the dorsal aorta take their origin from curious raised funnels which project into the lumen of the vessel (fig. 8c). Wagenvoort (1952) has reported two sphincters of smooth muscle that surround respectively the apex and base of the funnel and can close it. Their function is unknown. These funnels may perhaps ensure that the particular segmental vessel receives blood with an adequate content of erythrocytes. Axial streaming may result in the erythrocytes concentrating in the centre of the vessel and the plasma along the margin. Grodzinski (1964) has published photographs demonstrating very clearly the lamina flow in vessels of the trout yolk sac. Wagenvoort (1952) has reported experiments with hydraulic models which give some support for this suggestion.

In all fish the dorsal aorta sinks more deeply into the recess formed by downgrowths of the haemal arches, as it passes towards the back of the abdominal cavity. In the post-abdominal trunk the haemal canal is completed by the union of the haemal processes below it which enclose the dorsal aorta and caudal vein.

Each intercostal artery in *Heterodontus* (Birch, Carre and Satchell, 1969), *Carcharinus* (Satchell, 1965), and *Squalus* (Hanson, 1967) bears robust arterial valves close to its origin from the aorta (fig. 8d). They face away from the lumen of the aorta so that they would tend to close when pressure in the peripheral arteries rises due to compression of the vessels during swimming. These valves may help to ensure that the movement of blood in these vessels, caused by the waves of muscular contraction that pass along the trunk, is always from the dorsal aorta into the caudal vein. Valves are not present in the intercostal arteries of *Raja binoculata* and *Hydrolagus*

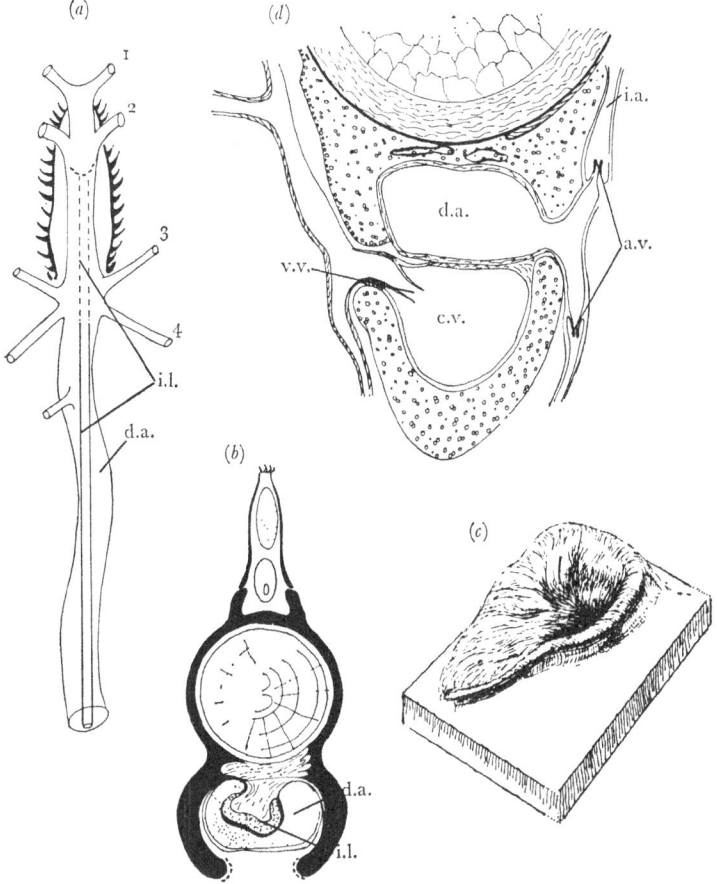

Fig. 8. (*a*) Dorsal aorta of *Clupea harengus*. (*b*) Transverse section of spine of *Clupea harengus* showing ligament in dorsal aorta. (*c*) Arterial funnel from dorsal aorta of *Petromyzon fluviatilis*. (*d*) Transverse section of haemal canal of *Heterodontus portusjacksoni* showing valves in intercostal arteries and veins. These have their origins half a segment apart; they are here shown at the same level. (a.v. = arteral valves, c.v. = caudal vein, d.a. = dorsal aorta, i.a. = intercostal artery, i.l. = internal ligament, v.v. = venous valves, 1, 2, 3, 4 = epibranchial arteries.) (*a, b* from De Kock and Symmons, 1959; *c* from Wagenvoort, 1952.)

colliei (Hanson, 1967); this may be associated with the reduction of the postpelvic trunk as a locomotory organ in these genera. *Raja* propels itself by undulatory movements of the pectoral fins; *Hydrolagus* rows itself along with its mobile pectoral fins. Both genera maintain the postpelvic trunk relatively still.

CHAPTER 3

The Veins

The structure of fish veins and the location of the venous valves

Much more is known of the anatomy, histology and physiology of the veins of elasmobranch than of teleost fish. Nevertheless it is clear that they differ. In the sharks and rays the veins are, apart from those of the hepatic portal system, either capacious sinuses or semirigid incollapsible tunnels in dense connective tissue. In the bony fish sinuses are much less common and the veins are discrete tubular vessels that can readily be dissected free from the surrounding tissue. In elasmobranch fish the olfactory, orbital, inferior jugular, anterior cardinal, posterior cardinal and gonadial sinuses are large blood-filled spaces in which the vein wall has disappeared except for the delicate endothelium which spreads over the surface of the adjacent organs. Each of these large venous reservoirs exhibits its own pattern of pulsatile pressure depending upon the movements of structures adjacent to it. In contrast, the lateral abdominal vein and the dorsal, lateral and ventral cutaneous veins are incollapsible tunnels in connective tissue in which the vein wall has undergone an intercrescence with the surrounding tissue. Indeed the only veins in elasmobranch fish which are free tubular vessels are those of the hepatic portal system. The classical studies of the elasmobranch venous system based on injected specimens have tended to ignore this significant feature.

In terrestrial vertebrates valves occur along the length of the long veins of the limbs and are commonly located just proximal to the entry of a tributary vein. In elasmobranch fish the main longitudinal vessels such as the caudal vein, lateral abdominal veins and cutaneous veins do not have valves along their length; valves do, however, occur in two situations. They are found (*a*) in segmental tributary vessels where these enter the main longitudinal veins, and (*b*) at the central ends of the longitudinal vessels where they enter the ductus cuvieri. Thus prominent valves occur at the central ends of the anterior cardinal sinus, lateral abdominal veins, and lateral cutaneous veins. But long veins such as the caudal vein, which extends from the pelvic bar to the tail, or the lateral cutaneous vein, which extends the entire length of the trunk, have no valves within their lumen (Birch et al. 1969).

Venous pressure

Venous pressures in fish are low and it is only with the advent of electrical manometers of high sensitivity that some reliance can be placed on reported values (table 2).

TABLE 2. *Venous pressures in elasmobranch fish in* cmH_2O

	Raja binoculata	Squalus suckleyi	Heterodontus portusjacksoni	Mustelus antarcticus
Caudal vein		2·6 – 2·4	2·1 ± 1·1	1·7 ± 1·2
Hepatic portal vein	2·9 – –1·4	2·5 – 2·3		
Hepatic vein	0·8 – –1·2	1·1 – 0·6		
Common cardinal vein	0·0 – –0·9	0·3 – –0·6		
Posterior cardinal sinus	0·5 – –0·5	0·2 – –0·6	–2·8 ± 1·7	
Lateral abdominal vein (central end)			–3·0 ± 2·0	
Anterior cardinal sinus		1·6 – 0·2		

Data for *Raja, Squalus* from Hanson (1967), for *Heterodontus* from Birch et al. (1969), for *Mustelus* from Satchell (1965).

The values quoted suggest that a gradient of pressure exists between the peripheral vessels remote from the heart, such as the caudal vein, where pressures are always positive, to vessels such as the cardinal sinuses and hepatic vein where they are often negative. All of these determinations of venous pressure are from elasmobranch fish. Much less is known of teleost fish. Mott (1950) reported pressures of −5·4 to 6·7 cmH_2O in the hepatic vein of an eel. Stevens and Randall (1967a) reported pressures of 9·4 to 13·5 cmH_2O in the subintestinal vein, a branch of the hepatic portal vein in *Salmo gairdneri*. These pressures approximately doubled during exercise.

The concept that venous return in elasmobranch fish is

effected by the residue of positive pressure remaining after the passage of blood through the peripheral capillary bed, i.e. the *vis a tergo*, and the sucking action of the heart, the *vis a fronte*, is borne out by pressure records along the length of the lateral cutaneous vein in *Heterodontus* (fig. 9). These records show that

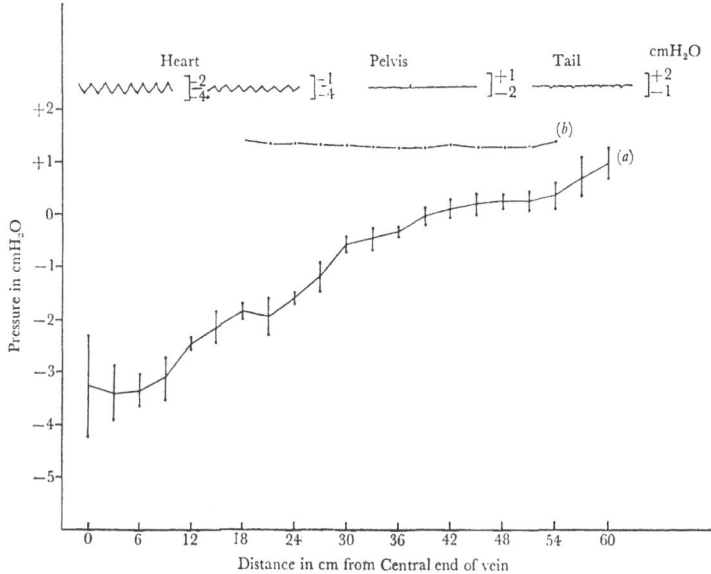

Fig. 9. Pressures along the length of the lateral cutaneous vein of *Heterodontus portusjacksoni* (a) at rest, (b) with central end of vein blocked. Vertical lines in (a) portray range of values in three successive determinations. The four traces inserted above the graph portray the fluctuations of venous pressure at each level, caused by cardiac suction. (After Birch *et al.*, 1969.)

there is a gradient along this vessel, from a pressure that is positive and steady at the peripheral end to a pressure that is negative and fluctuating in time with the heart beat at the central end (fig. 9a). Blocking the central end of this vein abolishes this gradient (fig. 9b) (Birch *et al.*, 1969). The caudal vein has the resistance of the renal portal circulation between it and the heart, and does not exhibit fluctuating pressures at rest. The large central sinuses that communicate with the sinus venosus do exhibit a fluctuating negative pressure derived from that of the atrium (fig. 2). Recordings of these have now been reported from the hepatic vein of *Squalus suckleyi* (Johansen and Hanson, 1967) and of *Raja binoculata* (Hanson, 1967), the posterior cardinal sinus of *Squalus suckleyi* (Hanson, 1967) and the central end of the lateral cutaneous vein in *Heterodontus* (Birch *et al.*, 1969). Birch *et al.* (1969) have recorded blood pressures in front of and behind the valve at the central end of the lateral cutaneous vein of *Heterodontus* and report that during diastole, when the pressure in the ductus cuvieri swings up towards zero, the pressure in the lateral cutaneous vein remains negative, suggesting that the valve at the end of the vein closes during diastole and opens during systole. Johansen *et. al.* (1968*a*) have shown that the velocity of blood flow along the vena cava of *Protopterus* is pulsatile, drops to zero at the end of diastole, and accelerates to a peak velocity as the central venous pressure swings negative. These studies suggest that in such fish venous inflow along some of the central veins halts as the valves close, and speeds again as they open and cardiac suction again lowers the pressure in these vessels. The cyclic opening and closing of venous valves effected by cardiac suction is a unique feature of the veins of fish; no valve in the veins of a mammal operates in this way.

The aspiration of the blood from the central veins by the heart possibly explains the two structural features of elasmobranch veins mentioned earlier. The sinuses provide a capacious reservoir from which the heart can readily withdraw blood. The semirigid vessels such as the lateral abdominal vein are able to withstand the negative pressure exerted on them without collapsing.

Auxiliary mechanisms of venous return

A   *The haemal canal*

The haemal canal in fish is a rigid and incollapsible tunnel below the spinal column that contains within it the dorsal aorta and caudal vein and serves to shelter these vessels from the waves of muscular tension that would otherwise bear on them as the sigmoid flexures of swimming pass down the trunk. The occurrence of valves in the intercostal arteries of the postpelvic trunk in *Heterodontus*, *Carcharinus* and *Squalus* has been noted in the previous chapter. These genera also possess paired valves in the intercostal veins (fig. 8*d*). They occur in the embouchments where the vessels enter the caudal vein,

34

and point towards the lumen of this vessel. They are absent in *Raja* and *Hydrolagus* (Hanson, 1967). Satchell (1965) suggested that the waves of muscular contraction compress the arteries, veins and capillaries of the trunk musculature, and that the orientation of both arterial and venous valves is such that blood cannot return to the dorsal aorta, but can pass only into the caudal vein. In the isolated post-pelvic trunk of *Heterodontus* perfused with dextran-saline through the dorsal aorta, swimming movements evoked by electrical stimulation of the spinal cord increased the outflow from the caudal vein by 46% above the resting level. Each lateral flexion of the body caused the expulsion of a jet of fluid from the caudal vein.

The haemal canal and its contained caudal vein, with its valved intercostal arteries and veins may be viewed as a mechanism whereby some of the power from the sigmoid waves of muscular contraction that serve to move the surrounding water backwards is deployed to move the blood in the caudal vein forwards. The importance of sheltering the caudal vein with its low venous pressure, from the waves of muscular compression that would otherwise squeeze the blood towards the tail, is apparent. It may well be true that the haemal arches sheltering the caudal vein are an essential part of this venous pump. They can certainly be detected in fossil fish as far back as the Devonian period.

B *The tail pump*

Mayer (1888) has described some of the relevant vascular anatomy of the tail of elasmobranch fish, and the paper by Birch *et al.* (1969) gives a more detailed description of that of *Heterodontus*. A band of muscle lies below the caudal extension of the spinal column; its fibres are inserted on to the radial cartilages and their contraction deflects the tail. Between the cartilages and the muscle are two longitudinal venous sinuses, separated by the blocks of muscle. The upper one communicates with the caudal vein by five or six valved vessels; the lower one in turn pours its blood into the upper sinus also by valved vessels (fig. 10). The two venous sinuses have valves along their length so arranged that blood entering them near the base or the tip of the tail will be directed towards the middle region, and from there into the caudal vein. The two sinuses receive blood from the dorsal, lateral and ventral cutaneous veins. Birch *et al.* (1969) report that pressure falls in these vessels as the tail moves and suggest that the mechanism constitutes a pump that transfers blood from the posterior ends of the cutaneous vessels into the caudal vein. In resting *Mustelus antarcticus* waves of contraction pass along the row of muscles that deflect the tail, squeezing the venous reservoirs that lie beneath them. The tail itself makes very little movement. These waves of muscular contraction occur continuously at rest and are dependent upon the presence of an intact central nervous system as they are activated by motoneurons in the spinal cord; they are paralysed by curare (Satchell, 1969).

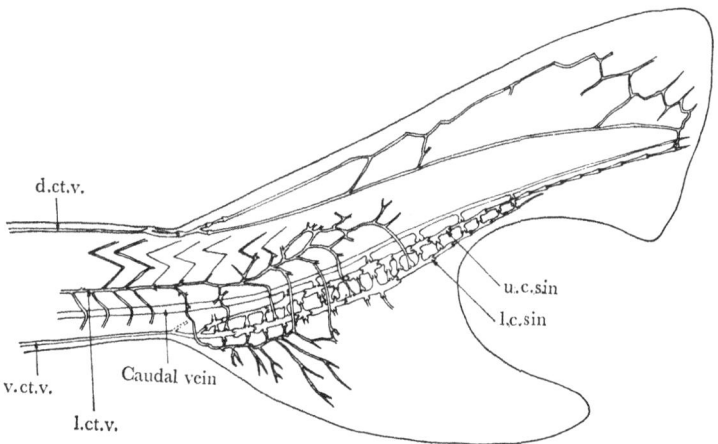

Fig. 10. The veins and their valves in the tail of *Heterodontus portusjacksoni*. (d.ct.v., l.ct.v., v.ct.v. = dorsal, lateral and ventral cutaneous veins; u.c.sin., l.c.sin. = upper and lower caudal sinus.) (After Birch *et al.*, 1969.)

This mechanism forms an interesting intermediate stage between that of *Heterodontus* and other sharks, where pumping is dependent upon tail movement, and the specialised lymph hearts of teleost fish and Amphibia which also are driven from motoneurons in the spinal cord. The caudal hearts of *Myxine* (Johansen, 1963) and of *Polistotrema* (Greene, 1899) also consist of paired venous sinuses compressed by the pressure of skeletal muscles against a median cartilage. They serve to pump blood from the large subcutaneous venous sinuses in the posterior trunk into the caudal vein, and Greene showed that their activity is dependent upon motoneurons in the posterior spinal cord, and is diminished during swimming. An examination of this region in other species is likely to be fruitful in demonstrating intermediate stages in the evolution of these accessory venous and lymphatic pumps in the tail.

c *The median fins*

The detailed vascular anatomy of the median fins has been described by Mayer (1888), by Marples (1935/36) in *Squatina*, and by Birch *et al.* (1969) in *Heterodontus*. The cutaneous vein bifurcates around the fin to form a vena circularis; at or near the bifurcation a vessel runs into the loose connective tissue at the base of the fin (fig. 11) and divides to form two small

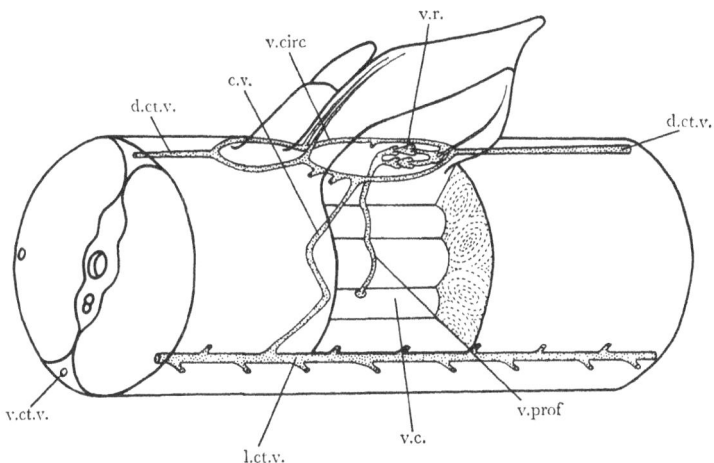

Fig. 11. The venous pump associated with the median fins of *Heterodontus portus-jacksoni*. (d.ct.v., l.ct.v., v.ct.v. = dorsal, lateral and ventral cutaneous veins; v.circ. = vena circularis; v.prof. = vena profunda. v.c. = caudal vein; c.v. = one of the large segmental vessels connecting the vena circularis with the lateral cutaneous vein; v.r. = venous reservoirs.) (After Birch *et al.*, 1969.)

venous reservoirs equipped with valves at both ends. They empty into the vena profunda which opens to the caudal vein. Birch *et al.* (1969) describe records of pressure from the dorsal cutaneous vein which suggest that fin movements pump blood from the cutaneous vessels to the caudal vein.

In the holocephalian fish, *Hydrolagus colliei*, Hanson (1967) reports that there is a large venous sinus below the single dorsal fin which communicates with the posterior cardinal sinus. Raising the fin aspirates blood from the cardinal sinus, and lowering it causes a rise in pressure in the cardinal sinus. The function of this fin pump requires further investigation.

D *Respiration and venous return*
The location of the anterior cardinal, hyoidean and inferior jugular sinuses in elasmobranch fish, is such that they are likely in some measure to be compressed during expiration and dilated during inspiration. Hughes and Ballintijn (1968) report that in *Scyliorhinus canicula* the constrictor muscles of the gill pouches contract in a peristaltic sequence which would be expected to move the blood down the anterior cardinal sinus towards the valved opening into the ductus cuvieri. Records from the posterior cardinal sinus of *Squalus* (Hanson, 1967) show small increases of pressure caused by respiration. However, no systematic study of the role of respiratory movements in venous return has yet been made in elasmobranch or teleost fish. In *Myxine* respiratory movement brings about a return of blood from the lacunar spaces to the veins.

CHAPTER 4

The Capillaries

Capillary beds represent specialised segments of the circulatory system in which the passive process of diffusion is facilitated by the provision of thin walls and a large surface area. This is achieved by the division of arterioles into numerous parallel channels of microscopic dimensions. The great increase in total cross-sectional area in the small vessels of the gill lamellae and tissues causes the velocity of flow to be reduced, and allows more time for diffusion to occur; no studies of this aspect of microcirculation in fish have yet been made.

The vasculature of the gill lamellae

On each side of a teleost fish are four gill arches (Hughes and Grimstone, 1965; Munshi and Singh, 1968), each of which bears a double row of gill filaments. In elasmobranch fish there are five to seven such rows and the two filaments are not free but are separated by and attached to a septum which partitions the gill cavity into five to seven separate pouches (Kempton, 1969). Each filament bears on its upper and lower surface, rows of closely packed leaf-like structures, the secondary lamellae. In *Gadus pollachius* they are approximately 10 μm thick; each consists of two epithelia, spaced apart by numerous pillar-cells. These pillar-cells are 3–4 μm in diameter and subdivide the cavity within the lamellae into numerous blood-filled channels; at their ends the pillar-cells extend laterally, their flanges abutting against those of adjacent pillar-cells so that all surfaces of the blood-filled channel are lined with pillar-cells or their extensions. The inner pillar-cell layer is separated from the outer epithelial layer by a basement membrane; from this columns pass across the lumen of the lamellae, each enfolded within the outer wall of the pillar-cell. Any one pillar-cell may include six to eight such columns. It is believed that the basement membrane and the columns are collagenous.

The pillar-cells have been likened to endothelial cells; the channels they line are analogous to capillaries. Blood enters these channels from branches of the afferent branchial arterioles and passes between the pillar-cells to be collected into the efferent branchial arterioles. The thousands of secondary lamellae are, like the alveoli of the lung, in parallel. Hughes (1966a) in a study of 14 species of marine teleost, found that there may be from 52–689,000 lamellae, with lengths ranging from 0·03–0·84 mm. Interposed between the afferent branchial arteries and the channels within the secondary lamellae is a column of vascular tissue, the cavernous body, consisting of blood-filled spaces lined with pigmented cells. Its function is unknown (Kempton, 1969).

The barrier across which the respiratory gases must diffuse is comparable in thickness to that of the mammalian lung. Hughes and Grimstone (1965) report that in *Gadus pollachius* it is 1–3 μm, comprised as follows: external epithelial layer 0·4–2·5 μm, basement membrane 0·3 μm, pillar-cell flange 0·1–0·3 μm. Calculations based on the diffusion coefficients of oxygen in connective tissue and water, the known area of lamellar surface, and the total oxygen consumption of fish, suggest that diffusion alone is sufficient to account for the supply of oxygen to the blood.

At the boundary of the gill lamella is a channel in which no pillar-cells are present to impede the flow of blood. The velocity of blood flow in this channel may be higher than in those situated nearer the base of the lamellae; this is probably also true of the velocity of water flowing in its counter-current pattern outside it (Hughes, 1966a). Steen and Kruysse (1964) suggest that in teleost fish not all of the blood that flows from the afferent to the efferent arteries passes through the respiratory lamellae, as there may be two other alternative channels. Blood may pass through either a central lymphatic compartment or through shunt vessels at the tips of the gill filaments. Adrenaline causes the blood to pass preferentially through the lamellar channels and increases the oxygenation of the blood of the eel. Hughes and Wright (1970) report that there are smooth muscle cells in the walls of the fine channels joining the efferent arterioles to the lamellar channels. If these dilate in response to adrenaline they could bring about this preferential perfusion of respiratory channels, and the decrease in gill vascular resistance that this agent is known to effect.

Hughes and Grimstone (1965) and Newstead (1967) report that pillar-cells contain numerous fibrous elements orientated longitudinally; if these were contractile they could pull the two epithelia of the lamella together and diminish the distance between the middle of the lamella and the water. The dimensions of the intralamellar vascular channels approximate to those of the erythrocytes of the species, and constriction of them may well deform the red cell. Guest et al. (1963) suggest that deformation of red cells has two effects. It brings a larger portion of the surface area of the red cell into proximity with the vessel wall which would enhance diffusion, by diminishing the distance to the middle of the cell. In addition it may, by its reorientation of the haemoglobin molecules within the cells, increase the oxygen affinity of the haemoglobin. Drake et al. (1963) have shown that deformation of the human red cell with lecithin resulted in a drop of the P_{50} (see p. 67) from 25–20 mmHg. The part that these various mechanisms may play in enhancing the transfer factor of the gills during hypoxia and exercise and in equating the conflicting demands of oxygenation and osmoregulation across the gill epithelium is still conjectural.

The peripheral capillaries

Our knowledge of the fine structure of the peripheral capillaries of fish is very meagre. Electronphotomicrographs of the capillaries in the electric organs of *Electrophorus electricus* and *Malapterurus electricus* show that there are two structural components interposed between the blood plasma and the tissue fluids; one of these is the endothelial cell and the other is a dense basement membrane that forms a continuous layer free from holes. The endothelial cells have dense attachment zones and their walls exhibit cavitations 20–60 nm across. The lumen of the capillaries is 6.5×3.5 μm in *Electrophorus* (Bennett et al., 1959).

Casley-Smith and Hart (1970) have studied the capillaries of the intestine, pancreas, kidney and ciliary body of *Heterodontus*. Electronphotomicrographs show that the venous limbs of the capillaries in all these tissues are fenestrated; the fenestrae may be open, or closed with a diaphragm. Moreover the capillaries have connective tissue fibres which pass from the abluminal membrane to surrounding connective tissue. Similar structures occur in mammalian lymphatic vessels; Casley-Smith and Hart suggest that in *Heterodontus* they serve to maintain the patency of the capillaries against the negative pressure present in the venous system.

The microcirculation of the skin of fishes has been extensively studied. Jakubowski (1963) has described the capillary network that lies in the dermis of *Cottus gobio*, *Rhombus maeoticus* and other species. In *Cottus* the capillary net is in the lower and middle dermal layers, while in *Rhombus*, *Pleuronectes* and *Lota* it is in the upper dermis. He emphasises that in fish with scales, irrespective of the extent of their development, the capillary net exists between the scale and the epidermis, and that the widespread opinion that scales cut off capillaries from the epidermis and render cutaneous respiration impossible, is without foundation.

A comparison of the degree of vascularisation of the skin of fish and Amphibia shows that species such as *Cyprinus carpio*, *Misgurnus fossilis* and *Nemachilus barbatulus* have as vascular skins as some Amphibia such as *Rana terrestris* and *Xenopus laevis*. When the length of skin capillary per gram of body weight is compared, the following data emerge: *Cyprinus carpio* 6·0 m, *Misgurnus fossilis* 5·9 m, *Nemachilus barbatulus* 5·8 m, *Rana terrestris* 6·1 m, *Xenopus laevis* 5·2 m. The highly vascular skin may be used for respiration in such species. Berg and Steen (1965) report that an eel breathing in air consumes about half as much oxygen at 7° C as it does in water and the gills are responsible for two-fifths of this amount. Using the data provided by Jakubowski (1960) on the vascularity of the eel skin, and the known diffusion constants of oxygen in air and muscle tissue they calculate that the demonstrated oxygen uptake is in accord with the anatomical data. They noted a hyperaemia of the dorsal fin during air breathing. Cutaneous respiration is undoubtedly of importance to eels; at 7° C an eel can stay in air for several weeks. Hughes (1966*b*, *c*) has suggested that cutaneous respiration may have been crucially important in the evolution of air breathing by the early vertebrates. The skin could augment both oxygen uptake and carbon dioxide elimination in the early stages when the evolving lung was insufficient for their needs.

Retial capillaries

Retial capillaries differ from true capillaries both in their anatomical relations and in their function. They are vessels of capillary dimensions located in the pathways of arteries and veins, on the way to and from the tissues. The arterial and venous capillaries are arranged in a parallel array; the counter-current pattern of blood flow that this permits, enables some agent to be concentrated at a particular site in the body. Three such systems have been the subject of recent investigations. The retia of the teleost swim-bladder concentrate oxygen, and sometimes nitrogen within the bladder so that when inflated, it provides buoyancy even at the great hydrostatic pressures of the ocean depths. The retia of tuna fish and Mako shark muscle concentrate heat within the interior of the body and enable such fish to be partially warm blooded. The retia in some teleost eyes concentrate oxygen within the optic chamber; this appears to be necessary for the retinal function of certain sight-orientated predacious fish.

The retia of the swim-bladder

The swim-bladder is primarily a hydrostatic organ that provides some teleost fish with neutral buoyancy. Bottom-dwelling fish that lack a swim-bladder are about 5% denser than seawater and have to swim constantly to avoid sinking (Denton, 1961). The total pressure of the gases within the bladder must be approximately equal to the hydrostatic pressure in the surrounding water; fish with swim-bladders occur down to depths of 7,000 m (Nielsen and Munk, 1964) and are common at depths of 300–1,000 m. These depths correspond to pressures of 20–100 atmospheres.

A conspicuous structure in the swim-bladder of such fish is the gas gland and the associated *rete mirabile*. It is supplied with arterial blood from the coeliac artery or a more posterior branch of the dorsal aorta; venous blood is returned to the hepatic portal vein (Harden-Jones and Marshall, 1953). The arterial blood passing to the gas gland traverses the arterial capillaries of the retia, which run in a closely packed parallel array with the venous capillaries returning blood from the gland (fig. 12a, b).

43

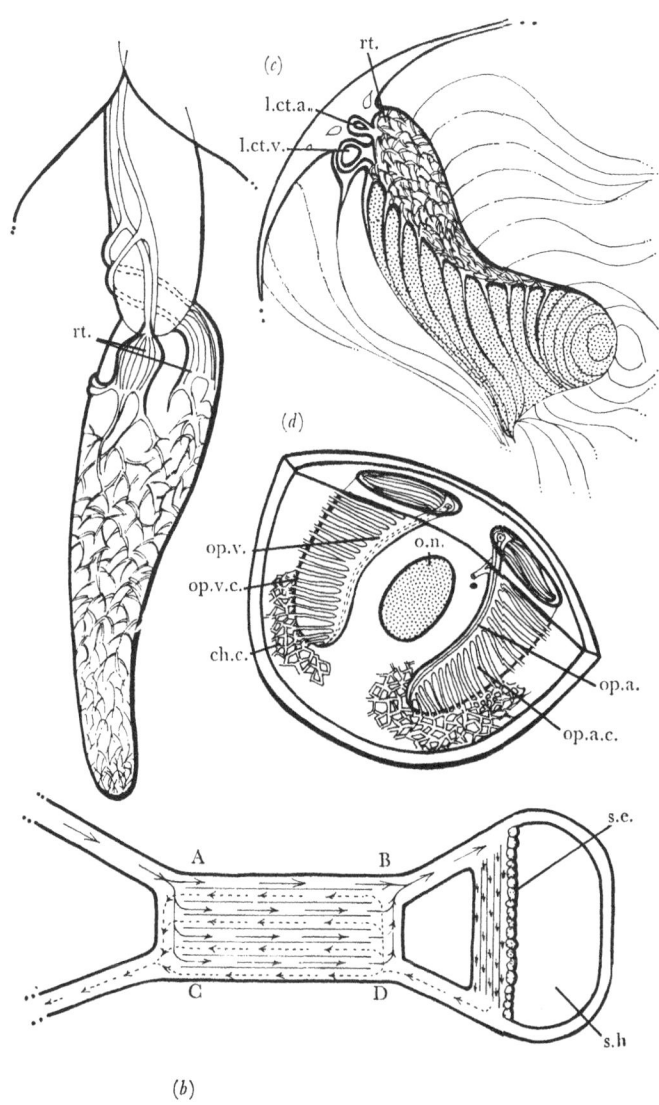

Fig. 12. Retial systems in fish. (*a*) The two retia of the swim-bladder of *Anguilla*.
(*b*) Diagram of the same showing direction of blood flow. (*c*) The trunk muscle
rete of *Isurus oxyrinchus*; the red muscle has been stippled. (*d*) The choroidal retia
of *Salmo gairdneri*. (A–B = arterial retial capillaries, C–D = venous retial capillaries, ch.c. = chorio-capillaries, l.ct.a. = lateral cutaneous artery, l.ct.v. = lateral
cutaneous vein, o.n. = optic nerve, op.a. = ophthalmic artery, op.v. = ophthalmic venous sinus, op.a.c. = arterial capillaries from ophthalmic artery which run
in parallel array with op.v.c., the venous capillaries leading to the ophthalmic
venous sinus, rt. = rete, s.b. = swim-bladder, s.e. = secretory epithelium.)
(*a, b* after Steen, 1963; *c* after Carey and Teal, 1969*a*; *d* after Barnett, 1951.)

44

Retial capillaries are longer than those of any other organ in the body; reported lengths are 1·2 cm in *Melamphaes* (Marshall, 1960), 1 cm in *Synaphobranchus* (Scholander, 1954), 0·4 cm in *Anguilla* (Krogh, 1959). These figures may be compared with the length of muscle capillaries in *Anguilla*, which are seldom more than 0·5 mm long. The retial capillaries of the eel are 9·5 × 7·1 μm in diameter, and have a cross sectional area of 71 μm². They are packed together in such a way that the maximum surface contact is achieved between them. Marshall (1960) reports that the capillaries may exhibit either hexagonal or mosaic packing. Scholander (1954) remarks that these are the only two solutions to the topological problem of making four polygons (black or white) meet at one point in such a way that black always borders white. Krogh (1959) reported that the two retia of an eel contained 88,000 venous and 116,000 arterial capillaries with a total length of 352 and 464 m respectively. The total surface area of the capillary walls was 106 cm², and 105 cm² ; this surface area was contained in a volume of 0·064 cm³, of which 66·7% was occupied with blood. The surface:volume ratio of this was calculated by Steen (1963) to be 1700 cm²/cm³ and may be compared with that of the human lung, 100 cm²/cm³. The wall thickness of 1·5 μm (Scholander, 1954) is roughly equivalent to that separating the blood in the capillaries of the lung from the alveolar air. We must conclude that the retia do indeed provide a large surface area for diffusion.

The gas gland consists of a glandular epithelium which may be single celled, multicellular or folded. In some genera such as *Perca* and *Sternopteryx*, the epithelium may contain giant cells. The capillaries of the gas gland lie below the epithelium and in genera in which it is folded such as *Opsanus*, the capillaries penetrate between the folds (Fänge and Wittenberg, 1958). In the bathypelagic fish *Vinciguerria*, the giant cells are particularly large, 100–150 μm long, and the capillaries pass through the interior of the cells (Marshall, 1960). The cytoplasm of cells surrounding the capillaries has been reported by many authors to stain differently from the rest of the cell.

The swim-bladders of shallow-water fish usually contain a mixture of gases with a high proportion of nitrogen, and Salmonid fish may contain 95% of nitrogen. Fish from depths of 400 m or more have oxygen as the principle component of the swim-bladder gas, and oxygen is the chief gas secreted into the swim-bladder when it is artificially emptied (Alexander, 1966).

It is now largely agreed that the deposition of oxygen in the swim-bladder is primarily dependent upon its liberation from the arterial blood brought to the gas gland. The gas gland tissue is known to convert glucose to lactic acid even at 1 atmosphere pressure of oxygen (Ball, Strittmatter and Cooper, 1955); venous blood leaving the retia is reported to have 60 mg% more lactic acid in it than the arterial blood (Kuhn, Moser and Kuhn, 1962). The lactic acid increases the P_{O_2} of the blood in the capillaries of the gas gland, and in the venous capillaries of the retia by three mechanisms. The rise in hydrogen ion concentration diminishes the oxygen capacity of the blood so that some of the haemoglobin cannot combine with oxygen. This is the Root effect and is known to occur in the blood of fish that possess a swim-bladder. It decreases the oxygen affinity of the blood; this is the Bohr effect; it is also a common property of fish blood (Harden-Jones and Marshall, 1953). It decreases the solubility of oxygen in the blood; this is the salting-out effect. Steen and Berg (1966) reported that an eel with an anaemia so severe that its blood contained no functional haemoglobin, had a swim-bladder containing 70% O_2. In such a fish oxygen deposition must be achieved solely by the salting-out effect; this mechanism would also seem to be the only one able to explain the deposition of nitrogen at pressures greater than those in the ambient water (Alexander, 1966).

These three mechanisms combine to increase the P_{O_2} of the blood in the capillaries of the gas gland and oxygen diffuses into the swim-bladder. As Scholander (1954) points out, at 100 atmospheres pressure of O_2, one volume of blood would dissolve four volumes of oxygen and the venous blood flowing away from the swim-bladder would rapidly empty it, but for the rete. It operates, it has been suggested, in the manner outlined in fig. 12b. Venous blood with a high P_{O_2} passes along the retial capillary at D and loses its oxygen by diffusion to the incoming blood in the arterial capillary at B. This will have already gained some oxygen by diffusion but the venous P_{O_2} is sufficiently high for further oxygen to enter the arterial blood. As the venous blood passes along the capillary towards C its P_{O_2} falls as its oxygen diffuses across the retial walls. But further diffusion is still possible as it is partnered by arterial blood which has a lower P_{O_2}, having not yet passed far along the arterial capillary. A gradient of oxygen pressure thus exists, it is suggested, along the whole length of the retial capillaries, and diffusion can continue throughout the passage of the blood along them. This is an example of a counter-current multiplier system in which the partial pressure of oxygen in the swim-bladder is raised by the successive addition of increments derived from the incoming blood, to a level greater than in any other part of the system. The multiplier function of the rete occurs because the ratio of content of oxygen to its partial pressure is successively reduced along the length of the arterial capillary. Alexander (1966) has critically reviewed the mathematical exposition of this hypothesis presented by Kuhn *et al.* (1963).

A difficulty in this explanation has recently been pointed out by Berg and Steen (1968). If the capillary walls are permeable to oxygen, and they must be if the counter-current multiplier system is to operate, they are likely to be permeable to hydrogen ions too. Indeed Steen (1963) has shown that the pH is equilibrated across the retial capillaries in the eel. It must therefore be concluded that as the blood passes along the arterial capillary, it receives both hydrogen ions and oxygen by diffusion. The loss of hydrogen ions from the venous blood might be expected to decrease the P_{O_2} by a 'Root-on' shift. This would reverse the gradient of partial pressure of oxygen between the venous and arterial capillaries and short circuit the counter-current diffusion. However, Forster and Steen (1968) have reported that the half time for the release of oxygen from oxygenated haemoglobin by acidification, i.e. the 'Root-off' shift, is 50 msec at 23° C, but the half time for the Root-on shift is 10–20 sec. This slow response of the blood to decreasing acidity results, we may suppose, in the venous blood continuing to have a high P_{O_2} as it passes along the retial capillaries. The system is moreover sensitive to the rate of flow. Rapid flow will enhance the gradient of oxygen partial pressure across the retial capillaries, as it provides less time for the Root-on shift to proceed towards completion. Slow flow would reduce the gradient of oxygen pressure and bring gas production to a standstill. The Root-off shift is dependent upon carbonic anhydrase and Forster and Steen (1968) have reported that acetazolamid prolongs the half time of the Root-off shift from 50 msec to 30 sec. This would largely abolish the difference in rate constants between the Root-on and Root-off shifts and destroy the basis on which the hairpin counter-current multiplier mechanism depends. Fänge (1953) has shown that when inhibitors of carbonic anhydrase are injected into *Ctenolabrus* the deposition of oxygen is reduced.

The retia of fish muscle

In the majority of fish the trunk muscles consist of a central core of poorly vascular white muscle, and a marginal filet of richly vascular red muscle that extends along the ventrolateral surface. Both are supplied from segmental arteries arising from the dorsal aorta.

Two totally unrelated groups of fish which have departed from this arrangement, are the lamnid sharks and tuna fish. The lamnid sharks, comprise such species as *Isurus oxyrinchus* the Mako shark, *Lamnia nassus* the porbeagle shark and *Carcharodon carcharias* the white pointer shark. The tuna fish include such notable species as *Thunnus thynnus*, *T. albacares*, and *Katsuwonus pelamis*. All of these are powerful swimmers capable of long-distance migrations. In all the red muscle is abundant and forms a longitudinal filet intucked into and surrounded by the white muscle (fig. 12c). The dorsal aorta is reduced and the red muscle obtains its arterial blood from a new vessel, the lateral cutaneous artery, that arises from the posterior gill arteries or the rostral end of the dorsal aorta and runs down the body at the level of the lateral line. The blood is returned not into the caudal or cardinal vein, but into a lateral cutaneous vein (Kishinouye, 1923). The blood has thus to pass to and from the muscle in a centripetal and centrifugal direction. It is this re-routing of the muscle vessels that permits the retia to be interposed between the muscle and the heart (Carey and Teal, 1969a, b). The arterial and venous retial capillaries thus come to have the blood flowing in them in opposite directions. Warm venous blood leaving an active muscle flows centrifugally towards the skin; heat passes across the walls of the retial capillaries and is carried in the arterial blood back to the muscles. This is an example of a counter-current system. It depends upon the passive diffusion of heat and merely

48

conserves the heat generated within the muscle. The muscle can never accumulate more heat than it can generate by its metabolism. The muscle retia cannot be a multiplier system as there is no way of reducing the specific heat of blood in its passage along the arterial capillary. The rete imposes a steep temperature gradient between the central core of red muscle and the outside. Carey and Teal (1969a) point out that 1 ml of blood carrying, when saturated, 0·2 ml of oxygen, would yield approximately 1 calorie which would warm the blood 1° C. As thermal diffusion is ten times as rapid as molecular diffusion, blood held in the respiratory lamellae sufficiently long to have become oxygenated will have cooled to water temperature. Interposing the high impedance to heat transport of a retial system between the site of heat generation and the site of oxygen uptake allows heat to accumulate in the muscle. Carey and Teal show that both in the Mako shark, *Isurus oxyrinchus*, and in the blue fin tuna *Thunnus thynnus*, the temperature of the interior muscle is substantially above water temperature. In cross-sections of such fish the isothermic lines centre around the dorso-medial part of the red muscle. The curve of body temperature against water temperature has a flattened zone between 10–30° C in *Thunnus thynnus*, and between 22–30° C in *Isurus oxyrinchus*, showing that such fish are not only warmer than the surrounding water but that they can within limits, regulate their temperature. In some way the return of heat to the muscle by the retia can be regulated, although the mechanism of this is not yet understood.

The choroidal retia of teleost fish

In the choroidal layer of the retina of certain teleost fish, and of the holostean *Amia calva*, a horseshoe-shaped structure encircles the head of the optic nerve; it was for a long time termed the choroid gland. Albers (1806) first recognised that it was a rete formed of the apposed arterial and venous capillaries that carry the blood to and from the chorio-capillaries. This constitutes a network of capillary vessels that extends in the choroidal layer of the retina. Barnett (1951) has given an account of the choroidal retia of nine species of teleost fish.

The ophthalmic artery enters the orbit above and behind the optic nerve. It divides to form two arms; from each arises the clusters of parallel twigs radiating towards the periphery of the retina. These are the arterial capillaries of the rete. The two main arterial branches lie within the two branches of the ophthalmic venous sinus (fig. 12d). Into this empty the numerous venous capillaries that accompany and lie parallel with the arterial capillaries. Each venous capillary is sandwiched between two arterial capillaries. In cross-sections the structure is like that of the rete of the swim-bladder; the two streams of blood are separated only by the thickness of the two capillary walls. All of the blood supply to the choroidal rete is derived from the ophthalmic artery; other structures such as the optic nerve, lentiform body and extraocular muscles are supplied from the retinal artery, a branch of the internal carotid artery.

It has been shown by Wittenberg and Wittenberg (1962) that the P_{O_2} of the vitreous humour of teleost fish may be very high. Species which have an exceptionally well-developed choroidal rete have the highest P_{O_2}, e.g. the remora, *Echeneis naverales*, 435–1,320 mmHg, and bluefish, *Pomatomus saltatrix*, 240–820 mmHg. Species with smaller retia have a lower P_{O_2}, e.g. puffer, *Sphaeroides maculatus*, 365–575 mmHg, and goggle-eye jack, *Trachurops crumenophthalmus*, 317–566 mmHg. Choroidal retia are absent in all elasmobranchs, and the three species investigated by Wittenberg and Wittenberg (1962) *Mustelus canis*, *Raja oscillata* and *Dasyatis centroura* all exhibited low partial pressures of oxygen below 30 mmHg. The bluefish and remora are both fast swimming, predacious, sight dependent fish in which the choroidal retia are so greatly developed that the rear of the eyes bulge and the rete is rotated 90° from its usual symmetrical position. It was in these fish that concentration ratios of oxygen from 2:1 to 8:1 were recorded.

The physiology of choroidal retia is at present largely unknown. Allen (1949) and Barnett (1951) noted that the ophthalmic artery which supplies all the blood to the choroidal retia is unique in arising from the pseudobranch. This glandular structure possesses characteristic acidophil cells, known to be rich in carbonic anhydrase (Leiner, 1938; Maetz, 1953; Hoffert, 1966). Pflugfelder (1952) reported that blinding one eye of *Hyphessobrycon* and *Lebistes* caused the pseudobranch of that side to regress to one-third or less of the unoperated side. Parry and Holliday (1960) have shown that removing the pseudobranch of *Salmo*, *Clupea* and *Gadus* caused the choroidal retia to degenerate.

Wittenberg and Wittenberg (1962) have suggested that the high intraocular pressure of oxygen is necessary to supply the respiratory needs of the relatively avascular retina. Perhaps the acidophil cells of the pseudobranch are in some way involved in retial function and the maintenance of oxygen pressure. This reciprocal relation between the pseudobranch and the eye is certainly in need of further investigation.

CHAPTER 5

The Blood

The blood of fish like that of other vertebrates consists of a suspension of erythrocytes and leucocytes in a solution of proteins and electrolytes which constitutes the plasma. Klontz *et al.* (1965) report that in *Salmo gairdneri* the anterior third of the kidney constitutes the most active site of haemopoiesis; the middle third of the kidney and the spleen also have this function. McKnight (1966) reports that in *Prosopium williamsoni*, the mountain white fish, leucocytes are derived from the interlobular tissue of the kidney, and erythrocytes predominantly from the spleen.

The haemocytoblast is considered to be the totipotential free stem cell that gives rise to all other blood cells (Klontz *et al.*, 1969). It is derived from a reticuloendothelial cell and is lymphoid in shape with a central nucleus. It gives rise to large and small haemoblasts. The large ones give rise to granulocytes and macrophages, the small ones give rise to erythrocytes, thrombocytes and lymphocytes. Klontz *et al.* (1969) have developed a nomenclature for the various stages in the maturation of erythrocytes and leucocytes that parallels that used in mammalian haematology yet recognises the absence of haemopoiesis in bone marrow in fish. Bleeding stimulates erythropoiesis in fish and Zanjani *et al.* (1969) have demonstrated the presence of a humoral agent, the erythropoiesis stimulating factor (E.S.F.) in *Trichogaster trichopterus*, the blue gourami. Plasma from bled fish contains E.S.F. and increases the incorporation of radio iron into the red cells of fish in which erythropoiesis has been inhibited by starvation. Plasma from normal fish possesses minimal E.S.F. activity.

The erythrocytes

Fish, like all vertebrates other than mammals have nucleated erythrocytes, which vary widely in shape and size from one species to another. Usually they are oval, e.g. the erythrocytes of *Dasyatis centrura* measure $19.7 \pm 0.38 \mu m \times 13.8 \pm 0.24 \mu m$ (Hartman and Lessler, 1964). Some species have smaller erythrocytes which are circular in outline; those of *Lepomis macrochirus* are $10.9 \mu m$ in diameter (Smith, Lewis and Kaplan, 1952). The mean erythrocyte volume (M.E.V.) may vary by a factor of four if we contrast the large oval cells of elasmobranch fish (table 3) with the smaller spherical cells of some teleost fish.

Wintrobe (1934) points out that there is an inverse relationship between the size and number of red cells in a species and this is roughly borne out by the data assembled in table 3. It is also seen if we compare the M.E.V. and erythrocyte count of the two species of *Ictalurus*, *I. nebulosus* and *I. punctatus*. The haematocrits of these two catfish are closely similar; *I. nebulosus* has fewer larger cells and *I. punctatus* has more numerous smaller cells (Haws and Goodnight, 1962).

Of particular interest are certain fish of Antarctic waters which have few or no erythrocytes. *Trematomus borchgrevinki* (table 3) has a count of $0.66-0.8 \times 10^6$ cells/mm^3 which is low for a teleost fish (Tyler, 1960). *Chaenocephalus aceratus* totally lacks erythrocytes; its blood has no absorption band for haemoglobin (Ruud, 1954). Some of the fast-swimming scombroid fish of the open ocean have high blood counts, e.g. *Scomberomorus maculatus* 3.2×10^6 cells/mm^3 (Becker *et al.*, 1958); Saunders (1966b) reports that *Acanthurus bahianus* has a count of 6.48×10^6 cells/mm^3 which exceeds the normal range for human blood ($4.6-6.2 \times 10^6$ cells/mm^3).

The red cell count is correlated with the haematocrit, haemoglobin content and oxygen capacity. In species such as *Salmo gairdneri* in which the first three of these have often been determined, they are found to have correlation coefficients of between 0.6 and 0.8. Predictions of red cell number and haemoglobin content from the haematocrit have standard errors around 10% (Houston and De Wilde, 1968a). The haematocrit of *Salmo gairdneri* is known to vary widely with temperature and nutrition; nevertheless, there are undoubtedly real differences between species. The elasmobranch fish mostly have haematocrits below 25% (table 3). There is, however, no very obvious relation between the evolutionary position of a fish and the haematocrit of its blood, as is shown by the value for *Petromyzon marinus* of 33% (Thorson, 1959). Teleost fish

53

TABLE 3. *Some properties of the blood of fishes*

Species	Temperature (°C)	Mean erythrocyte Volume in μm^3	Diameter in μm	Erythrocyte count ($\times 10^6$ cells/mm³)	Haematocrit (%)	Haemoglobin concentration (g/100 ml)	Oxygen capacity whole blood (vol %)	Authors
Raja oscillata	—	—	—	0·2	20	—	6·0	Dill *et al.* 1932
R. erinacea	—	642–910 μm^3		0·07	—	—	—	Wintrobe, 1934
Squalus acanthias	—	650–1010 μm^3		0·09	18·2	—	—	Thorson, 1958 Wintrobe, 1934
S. suckleyi	—	—	—	—	13–26	—	3·5–4·5	Martin, 1950 Lenfant and Johansen, 1966
Trematomus borchgrevinki	−1·5		11·0 μm	0·66–0·80	26–32	3·5–4·0	6·1–7·0	Griggs, 1967 Tyler, 1960
Chaenocephalus aceratus	+2 − −1·7		—	0	0	0	0·45–1·08	Ruud, 1954
Ictalurus nebulosus			13·8±0·9 × 9·4±0·7 μm	1·22±0·04	27·9±6·5	6·9±0·4	8·85±2·96	Haws and Goodnight, 1962
I. punctatus			11·0±0·7 × 7·6±0·6 μm	2·16±0·06	29·4±4·9	6·6±0·3	8·97±0·5	Haws and Goodnight, 1962

Species							References
Cyprinus carpio	17	$186 \pm 7 \ \mu m^3$	$1 \cdot 43 \pm 0 \cdot 13$	$27 \cdot 1 \pm 2 \cdot 0$	$6 \cdot 4 \pm 0 \cdot 4$	$12 \cdot 5$	Houston and De Wilde, 1968b; Black, 1940
Carassius auratus	20–26	$163 \pm 30 \ \mu m^3$	$2 \cdot 12 \pm 0 \cdot 02$	$29 \cdot 6 \pm 5 \cdot 9$	$9 \cdot 8$	$12 \cdot 29$ "	Anthony, 1961; Falkner and Houston, 1966; Prosser et al. 1957
Salmo gairdneri	3–21	$252 \ \mu m^3$	$1 \cdot 29 \pm 0 \cdot 02$	$31 \cdot 58 \pm 0 \cdot 27$	$7 \cdot 42 \pm 0 \cdot 15$	9–10	Houston and De Wilde, 1968a; De Wilde and Houston, 1967
Salvelinus fontinalis	2	$294 \pm 6 \ \mu m^3$	$1 \cdot 23 \pm 0 \cdot 03$	$34 \cdot 1 \pm 0 \cdot 9$	$7 \cdot 2 \pm 0 \cdot 2$	11–$13 \cdot 9$	Holeton and Randall 1967b; Houston and De Wilde, 1969
Prosopium williamsoni	—	$13 \cdot 2 \times 9 \cdot 5 \ \mu m$	$1 \cdot 01 - 2 \cdot 34$	$28 \cdot 5 - 62 \cdot 2$	$7 \cdot 0 - 13 \cdot 7$	—	McKnight, 1966
Trichogaster trichopterus	—	—	$3 \cdot 06 \pm 0 \cdot 07$	$27 \cdot 3 \pm 1 \cdot 47$	$10 \cdot 42 \pm 0 \cdot 36$	—	Zanjani et al., 1969
Thunnus thynnus	—	—	$2 \cdot 3$	$41 \cdot 0$	—	—	Becker et al., 1958
Euthynnus lineatus	—	—	—	—	—	$16 \cdot 9 - 19 \cdot 9$	Klawe et al., 1963
Auxis rochei	—	—	—	—	—	$17 \cdot 8 - 21 \cdot 2$	Klawe et al., 1963

5

55

usually have haematocrits between 20–30% (table 3) but some have higher values. *Prosopium williamsoni* (28·5–62·2% (McKnight, 1966)), and *Thunnus thynnus* (41% (Becker *et al.*, 1958)), overlap the range for human blood of 38–53%.

The haemoglobin concentration of the blood of most teleost fish is between 7–12 g% (table 3); that of the Antarctic nototheniid fish *Trematomus borchgrevinki* is only 3·5–4·0 g% (Tyler, 1960). Tuna blood is rich in haemoglobin and Barrett and Connor (1962) report that skipjack, *Katsuwonus pelamis*, have a haemoglobin content of 16·7 g%. Klawe, Barrett and Klawe, (1963) report even higher levels of 17·8–21·2 g% in blue skipjack, *Euthunnus lineatus*. Again, these values equal or exceed the 12·0–17·5 g% regarded as normal for man.

The oxygen capacity of fish blood comprises oxygen in physical solution and oxygen in combination with haemoglobin. In *Chaenocephalus aceratus* the value of 0·45–1·08 vols% must comprise only the oxygen in solution as the species lacks haemoglobin (Ruud, 1954). Other Antarctic fish have higher values, e.g. *Notothenia coriiceps* 7·5–8·8 vols% (Hemmingsen, Douglas and Grigg, 1969). Elasmobranch fish have bloods of rather low oxygen capacity, e.g. *Squalus suckleyi* 3·5–4·5 vols% (Lenfant and Johansen, 1966). In carp and in *Salvelinus fontinalis* acclimation to high temperatures increases the red cell count, haematocrit, haemoglobin and oxygen capacity (Houston and De Wilde, 1968*b*, 1969). The response of *Ictalurus nebulosus* is the opposite of this; acclimation to higher temperatures causes a reduction in the haematocrit and oxygen capacity (Grigg, 1969).

The leucocytes

The three main categories of leucocytes found in mammalian blood, the granulocytes, monocytes and lymphocytes, have all been recorded in the blood of fish. Klontz (1969) however, denies that monocytes occur in the blood of *Salmo gairdneri* and believes that cells previously identified as monocytes are in reality circulating macrophages. A fourth category of white cell, the thrombocyte, occurs in the blood of fish as it does in that of other lower vertebrates; it is believed to be the source of clotting factors, and to replace the platelets of mammals in this regard.

The granulocytes include the three well-recognized classes of neutrophils, eosinophils and basophils. Neutrophils are much the most common and occur in the blood of many species of fish (Saunders, 1966*a*, *b*, 1967, 1968). Juvenile, band and segmented forms occur; the latter is the mature form. Eosinophils are often absent, but were present in 65 of 116 species of teleost fish and in each specimen of five species of elasmobranch fish examined by Saunders (1966*b*). Their sporadic occurrence Saunders suggests, may be a response to infection by worms; eosinophilia occurs as a response to parasitic infections in man. Basophils have been seen only rarely in fish blood; they occur in certain species such as *Synodus intermedius* and *Holocentrus ascensionis*. They have not been observed in elasmobranch blood and may not occur there. Klontz *et al.* (1969) suggest that in *Salmo gairdneri* both eosinophils and basophils are not derived from the progranulocyte, but are histogenous in origin and arise outside the vascular system in the connective tissue of the intestine. Fänge (1968) reports that in elasmobranch fish eosinophilic granulocytes are formed in a specialised mass of lymphomyeloid tissue, the organ of Leydig, in the wall of the oesophagus. Similar tissue occurs in the orbit and the base of the cranium in *Chimaera*, within the meninges surrounding the brain in *Amia* and *Lepisosteus*, and in the mesenteries of *Labrus*.

Some elasmobranch fish possess a fourth class of granulocyte, the heterophil, which is characterised by the presence of eosinophilic rodlike granules in the cytoplasm. Heterophils occur as alternative cells to neutrophils, the two types of cells not being present together (Saunders, 1966*a*).

Lymphocytes have been reported in the blood of many species of fish (Reznikoff and Reznikoff, 1934; Lieb, Slane and Wilber, 1953; Becker *et al.*, 1958). From the prolymphocytes, small and large lymphocytes differentiate. These are pleuropotential and may differentiate further into macrophages, plasma cells and thrombocytes. Small lymphocytes are 7–10 μm in diameter and are believed to be concerned in protein production (Klontz *et al.*, 1969). Macrophages are phagocytic and perform the task of removing exogenous and endogenous debris; they are cleared from the circulatory system in the anterior kidney in *Salmo gairdneri* (Klontz, Yasutake and Parisot, 1965). In response to a cutaneous injection of a pathogenic virus, macrophages infiltrate below the injection site, and there is a subsequent increase in circulating macrophages. Large lymphocytes too are involved in the chronic inflammatory response and ingest debris. Plasma cells are small ovoid leucocytes with eccentric nuclei which have been much studied because of their possible role in antibody production. In rainbow trout the relative numbers of plasma cells are unrelated to the level of circulating antibodies and the role of these cells remains obscure (Klontz, 1969).

Thrombocytes are the most common type of white cell and comprise approximately half the leucocytes in all of the 121 species of fish examined by Saunders (1966b). They are the biochemical equivalent of platelets in mammalian blood, providing a factor which converts prothrombin to thrombin. The thrombocytes of teleost fish are more fragile than those of elasmobranch fish (Doolittle and Surgenor, 1962). Thrombocytes are responsible for clot retraction in *Mustelus* (Doolittle, 1963).

The plasma proteins

The plasma proteins are responsible for contributing to the osmolarity of the plasma in which the blood cells are transported. They are the source of the colloid osmotic pressure which, by opposing the hydrostatic pressure of the blood, regulates the movement of water across the capillary wall. They supply the protein for the nutritional requirements of the tissues. They contribute to the buffering power of the blood and thus play a part in the regulation of pH. They are concerned in the defence of the body against injury and attack by pathogenic organisms. They serve to transport vitamins, hormones and certain inorganic ions such as iron.

Information about the plasma proteins of fish is fragmentary and scattered and some of the data available in the literature were derived when analytical methods were less reliable. It is clear that the total plasma protein is very variable ranging from 1·68 g% in *Cynoscion arenarius* to 6·19 g% in *Sciaenops ocellata* (Sulya, Box and Gunter, 1960). This latter figure approaches the 7 g% of human blood. The ray *Raja kenojci* and the shark, *Heterodontus japonicus* lack serum albumin (Irisawa and Irisawa, 1954) as do three species of Lamniform shark reported by Sulya, Box and Gunter (1961). These authors suggest that all elasmobranch fish lack serum albumin. In man three fourths of the colloid osmotic pressure of the blood is contributed by the albumin, and the absence of this fraction in elasmobranch blood suggests that the colloid osmotic pressure at the capillary membrane is very low. Albumin is certainly present in the blood of the advanced teleosts; Field, Elvehjem and Chancey (1943) reported that the total plasma protein of carp was 4·15 g%; of this albumin accounted for 2·82 g, globulin 0·79 g and fibrinogen 0·23 g. Many reports on the albumin: globulin ratio of fish serum show it to be lower than that of mammals. Lepkovsky (1929) quotes values of 1·28–2·08 in *Brevoortia tyrannus*, and 0·23–0·39 in *Squalus acanthias*. Flemming (1958) identifies four factors separable by electrophoresis in carp serum; albumin plus alpha$_1$ globulin, which migrate together, alpha$_2$, beta and gamma globulin. These four fractions comprise respectively 31·6–37·9%, 16·3–25·0%, 21·0–27·0% and 17·1–27·4% of the serum proteins of carp.

Klontz *et al.* (1965) reported an electrophoretic study of the serum proteins of *Salmo gairdneri* and recognised albumin, alpha$_1$, alpha$_2$, beta$_1$, beta$_2$ and gamma globulins. They noted, however, that none of these six major serum components can be correlated directly with any of the mammalian serum proteins. Inoculation of rainbow trout with Sacramento River chinook disease virus caused an elevation of the beta$_2$ globulins coincident with marked lymphoid hyperplasia. This suggests that in fish antibody protein is associated with beta$_2$ rather than with the gamma globulin fraction as it is in mammals.

The belief that electrophoretic patterns of plasma proteins may provide information of taxonomic value has led many workers to publish analyses of this type. Deutsche and McShan (1949) reported 12 separate components in the electrophorogram of the plasma of *Perca flavescens*. Serum transferrins have been identified autoradiographically by their ability to transport iron[59] (Moller and Naevdal, 1965; Barrett and Tsuyuki, 1967). Haptoglobins, which are specific proteins with the capacity to combine with haemoglobin, and occur associated with the alpha$_2$ globulin fraction in man, also occur in small quantities in *Catostomus insignis* and certain other catostomid fish (Koehn, 1966). There have, however, been no studies of the role of serum transferrins and haptoglobins in fish; we know only of their occurrence. Sano (1960) has shown that the levels of the alpha and beta globulins increase gradually with the growth of fingerlings of *Salmo gairdneri*, and that in females the beta globulin component reaches a maximum with the development of the ovary in September. Injection of *Oncorhynchus nerka* with oestradiol monobenzoate causes differences between male and female serum protein levels. These observations and many others reported in a review by Booke (1964) suggest that much remains to be learnt of the functions of the plasma proteins in fish.

The properties of fish haemoglobins

The haemoglobin of fish, like that of other vertebrates consists of a colourless protein, globin attached to an iron-containing pigment, haem, that imparts the red colour. This pigment consists of an isomer of protoporphyrin, comprising four pyrrole rings with an iron atom in the centre. This iron atom is hexavalent; four of the valencies are attached to the pyrrole rings, a fifth attaches the porphyrin to the protein and the sixth is available for combination with oxygen. Mammalian haemoglobins consist of four polypeptide chains, with a haem group attached to each. The four chains consist of two identical pairs, the alpha and beta chains. Beta chains consist of 146 aminoacid residues and have a very similar structure to mammalian myoglobin. Alpha chains consist of 141 residues; they resemble the beta chains in many of their sequences but have a shortcut across one of the loops (Perutz, 1964). Each chain consists of helical and non-helical segments. The chains coil around each other and enclose a central water-filled space; the four haem groups are each enfolded into one of the chains so that the valencies which will bind oxygen stick outwards. The contacts between unlike chains are chiefly non-polar; the contacts between like chains are polar. The molecular weights of these tetrameric haemoglobins are around 64,000.

The overall secondary, tertiary and quarternary structures of the haemoglobin from a variety of higher vertebrates have been shown to be very alike despite the presence of different aminoacid residues in certain positions in the chains. This is because the overall configuration of the haemoglobin molecule is primarily dependent upon the strategic location of certain polar and non-polar groupings. These similarities give us confidence that, although our knowledge of fish haemoglobins is only now emerging into the literature, they are, at least in the elasmobranch and teleost fish, similar in molecular shape though not identical in aminoacid sequence to those of higher vertebrates. Sick, Frydenberg and Nielsen (1963) report that the molecular weight of the haemoglobin of plaice, *Pleuronectes platessa*, is approximately 70,000 suggesting that it is a tetramer. Grigg (1969) drew the same conclusion from sedimentation studies of the haemoglobin of *Ictalurus nebulosus*. Braunitzer and Hilse (1963) have studied the aminoacid sequence of the alpha chains of carp haemoglobin. Many of the sequences are identical with those of human haemoglobin though the gap between residues 46–47 is closed. De Marco and Antonini (1958) have obtained the haemoglobins of *Pelamys sarda* and *Thunnus thynnus* in crystalline form and reported on their aminoacid composition.

The Cyclostomata stand apart from all other vertebrates in possessing a monomeric haemoglobin with only one chain; that of *Petromyzon marinus* has a molecular weight of $18,200 \pm 400$ (Lenhert, Lowe and Carlson, 1956). It resembles in some respects the myoglobin of the muscle of higher vertebrates.

Multiple and hybrid haemoglobins

Many fish are known to possess multiple haemoglobins which differ in aminoacid composition, electrophoretic mobility and solubility. Some species have three different haemoglobins (Buhler and Shanks, 1959; Chandrasekhar, 1959). *Oncorhynchus keta* has two haemoglobins, termed slow (S) and fast (F); they are alike in possessing isoleucine, usually absent from mammalian haemoglobin, but they differ in other respects. The S component has more alanine than F, but less histidine, arginine, glutamic acid and only one-third to one-half as much cystine (Eguchi, Hashimoto and Matsuura, 1960).

Manwell and Childers (1963) report that the cross of the centrarchid fishes *Chaenobryttus gulosus* with *Lepomis cyanellus* yields a hybrid in which the blood contains 40% of a new haemoglobin made up of the alpha chains of one parent and the beta chains of the other. This hybrid haemoglobin has a greater affinity for oxygen than that of either parent. The naturally occurring North Sea hybrids between plaice and flounder have a hybrid haemoglobin (Sick *et al.*, 1963).

CHAPTER 6

The Transport of the Respiratory Gases within the Body

If sufficient oxygen is to be transported to the tissues, the various components of the circulatory system must achieve at least a certain minimum performance. There must be sufficient cardiac output, i.e. the myocardium must generate a sufficient pressure of blood to overcome the resistance of the peripheral vessels. There must at all times be an adequate volume of blood within the circulatory system to match the demands of the heart if the capacity of the system is increased by vasodilation. The blood must have a sufficient oxygen capacity. The haemoglobin must possess an appropriate oxygen affinity such that it can take up oxygen at the gills and unload it in the tissues, at the pressures which actually prevail there.

Cardiac output

Published values of cardiac output in fish vary widely. Some of this variation may be due to the use of different techniques. The accuracy of the Fick method may be compromised by uncertainties concerning the oxygen dissociation curve. The electromagnetic flowmeter provides a direct measurement of the velocity of flow but requires surgery to insert the probe. Few workers have been able to achieve anything approximating to a basal metabolic rate before their determinations. Some recently reported values are presented in table 4. The data suggest that the cardiac output often lies between values of 20–40 ml/kg/min; such values are less than half those reported for mammals, e.g. rat 260, goat 130, horse 70 ml/kg/min.

Blood pressure

Previous reviews (Mott, 1957; Johansen and Martin, 1965) have presented the earlier data on fish arterial blood pressures.

TABLE 4. *Cardiac output in fish*

Species	Temperature (°C)	Cardiac output (ml/min/kg)	Stroke vol. (ml/kg)	Method used	Author
Hydrolagus colliei	9±1	21·0	0·7	Electromagnetic flowmeter (EMF)	Hanson (1967)
Scyliorhinus stellaris	10±1	22·0	—	Fick principle	Baumgarten and Piiper (1969)
Squalus suckleyi	10±2	8–24·0	0·4–1·7	EMF, Fick	Hanson (1967)
Raja binoculata	10±2	10·0	0·7	EMF	Hanson (1967)
Squalus acanthias	11–17	9–33·0	0·25–0·93	Ligations, flow from cut ventral aorta	Burger and Bradley (1951)
Gadus morhua	—	9·0	0·31	EMF	Johansen (1962)
Myoxocephalus scorpius	15–18	21–34·0	—	Fick principle	Goldstein *et al.* (1964)
Salmo gairdneri	12–18	65–100·0	0·85–2·0	Fick principle	Holeton and Randall (1967b)
Ophiodon elongatus	12	58·0	—	Fick principle	Randall (1968)

TABLE 5. *Arterial blood pressures in fish in cmH$_2$O*

Species	Before gills	After gills	Difference	Percentage reduction of average pressure	Temperature (°C)	Author
Myxine glutinosa	6.5	1.3	5.2	80	—	Johansen (1960)
Hydrolagus colliei	28–19	17–14	11.5	34	9±1	Hanson (1967)
Mustelus canis	35–26					Sudak (1965a)
Squalus suckleyi	39–25	29–20	10.5	23	10±2	Hanson (1967)
Heterodontus portusjacksoni	44–30	28–22	16.8	32	18	Satchell (1969)
Raja binoculata	38–26	29–22	9.4	20	10±2	Hanson (1967)
Oncorhynchus tshawytscha	110.0	67.0	43.0	39	16±18	Robertson et al. (1966) *
Oncorhynchus tshawytscha	101.0	69.0	32.0	32	—	Greene (1904)
O. nerka	—	62–54	—	—	—	Smith et al. (1967)
Salmo gairdneri	53–43	39–34	15.9	25	10–12	Stevens and Randall (1967a)
Opsanus tau	—	18.2	—	—	18	Lahlou et al. (1969)

* Pressures recorded with fish briefly out of water; possibly this value is too high.

In table 5 are included only the more recent determinations, plus that of Greene (1904). There still remains a paucity of records for teleost fish.

The data would seem to suggest that teleost fish have higher blood pressures than elasmobranch fish. Uncertainty is caused because four of the five records of teleost fish are from the family Salmonidae, which are both active, and live for at least part of their life in fresh water. These high pressures may reflect merely the necessity of achieving the high glomerular filtration rate that accompanies life in fresh water, and may not be characteristic of the bony fish generally.

The loss of pressure across the gill vasculature amounts to one-quarter to one-third of that generated by the heart. The resistance of the gill vessels and the compliance of the ventral aorta combine to diminish the pulse pressure so that it is proportionately less in the dorsal aorta. The existence of a pressure pulse in the dorsal aorta suggests that the flow of blood through the gill vessels is pulsatile. Randall *et al.* (1969) report that flow in the efferent branchial arteries of the cod, *Gadus, morhua,* is pulsatile though less so than in the ventral aorta.

Blood volume

There have been numerous determinations of the blood volume of fish and Mott (1957) gives a selection of values derived from the earlier literature. Table 6 presents a spectrum of values

TABLE 6. *The blood volumes of various species of fish (volumes %)*

Species	Blood volume	Author
Petromyzon marinus	8·5	Thorson (1959)
Hydrolagus colliei	5·2	Thorson (1958)
Raja binoculata	8·0	Thorson (1958)
Squalus acanthias	6·8	Thorson (1958)
Polyodon spathula	3·5	Thorson (1961)
Mycteroperca tigris	3·3	Thorson (1961)
Epinephelus striatus	2·6	Thorson (1961)
Salvelinus fontinalis	5·7	Houston and De Wilde (1969)
Salmo gairdneri	3·5	Conte *et al.* (1963)
Salmo gairdneri	6·9 ± 1·8	Smith (1966)
Oncorhynchus nerka	5·4 ± 0·8	Smith (1966)
O. kisutch	6·1 ± 1·6–7·2 ± 0·1	Smith and Bell (1964)

determined by dye dilution techniques. Smith and Bell (1964) suggest that the complete admixture of the injected dye with the venous blood is a very slow process and that by allowing 4–5 hr to elapse before sampling, higher values of up to 6 vols% are obtained. Martin (1950) reported that the blood volume of pregnant *Squalus suckleyi* is increased from 5 to 10–13 vols%, with no change in the haematocrit. Houston and De Wilde (1969) found that acclimation to high temperature increased the blood volume of *Salvelinus fontinalis*. Fish acclimated to 20° C have a 25% greater blood volume than fish acclimated to 2° C.

The data from table 6 suggests that there may be a phylogenetic decrease in blood volume. This may in part reflect the increasing perfection of the venous system wherein the large venous sinuses of Cyclostomata and Elasmobranchii are replaced by narrow tubular veins in Teleostei. Stevens (1968) has determined the instantaneous blood volume of the separate tissues in *Salmo gairdneri*, at rest and after exercise, by injecting radio iodinated serum albumin and counting the emanations. He reports that the white muscle which constitutes 66% of the weight of the fish, contains only 15% of the blood. If the blood volume of the remaining one-third of the fish which is not white muscle, is separately calculated assuming it to have 85% of the blood, a value of 13 vols% is obtained. This suggests that the low blood volume of fish reflects the dominance of this mass of poorly vascularised white muscle.

Stevens also reported that there is relatively little change in the volume of blood in the muscle following exercise. Cardiac output is known to increase fivefold. This suggests that at rest many capillaries have the blood within them stationary and that the relaxation of arterioles during exercise, permits an increase in the velocity of flow through the capillaries without any notable increase in their capacity.

The oxygenation of haemoglobin

Haemoglobin has the ability to combine reversibly with oxygen when the oxygen pressure is high enough and to unload this oxygen as the pressure falls. This binding with oxygen involves a change from an ionic to a covalent bond without a change in valency of the iron. The reversibility of oxygenation is a property imparted to the haem by its association with the polypeptide chain. It involves configurational change in the haemoglobin molecule. The binding of oxygen to the haem causes the two beta chains to move closer together so that the distance between the iron atoms diminishes from 4·03 nm to 3·36 nm. Perutz (1964) has described the haemoglobin as a molecular lung and we can picture the molecule undergoing a respiratory motion as it passes in turn through the gills and the tissues.

When the partial pressure of oxygen in the blood is plotted against the amount of oxygen bound to the haemoglobin, the oxygen dissociation curve is derived. Such curves have been determined for many species of fish and are always non-linear. The properties of the blood of a species which are significant in terms of its ability to transport oxygen, can be assessed by examining three features of the oxygen dissociation curve:

(1) The shape of the curve, i.e. whether it is sigmoid with a flattened segment at low partial pressures of oxygen, or is hyperbolic.

(2) The position of the curve on the x-axis, i.e. the range of P_{O_2} within which the haemoglobin reversibly combines with oxygen.

(3) The effect that changes of pH and other ions have on the shape and range of the curve.

Hill (1910) developed an equation on the supposition that binding oxygen depends upon sub-unit interaction:

$$Y = (Kp^n)/(1 + Kp^n)$$
$$y = \text{degree of saturation,}$$
$$p = \text{partial pressure of oxygen,}$$
$$K \text{ and } n = \text{constants.}$$

The term n is an index of sub-unit interaction and curves in which $n = 1$ are hyperbolic. If sub-unit interaction results in an acceleration of oxygen binding as the pressure of oxygen increases, n will exceed 1 and the curve will tend to be sigmoid.

The oxygen affinity of a haemoglobin is commonly expressed as the P_{50}, i.e. the partial pressure of oxygen at which half of the haemoglobin is saturated.

A *Sigmoid and hyperbolic dissociation curves*

Because the binding of oxygen to the haem groups alters the configuration of the haemoglobin molecule, the affinity of the remaining haem groups may be altered; it may either be enhanced or depressed. These haem–haem interactions may thus serve to increase the affinity of the haemoglobin for oxygen when a certain critical level of saturation has been exceeded, and a sigmoid curve results. In mammalian haemoglobin the configurational changes between the alpha and beta chains greatly increase the affinity for oxygen of the fourth haem group when the other three have bound oxygen (Perutz, 1964). The absence of such haem–haem interactions will cause the dissociation curve to be hyperbolic. The monomeric haemoglobins of hagfish, *Polistotrema stouti* (Manwell, 1958a), and of lamprey, *Petromyzon marinus* (Wald and Riggs, 1951) are necessarily hyperbolic, with an *n* value of 1. The bloods of *Hydrolagus colliei* (Manwell, 1964) and *Squalus suckleyi* (Lenfant and Johansen, 1966) have a hyperbolic oxygen dissociation curve, as has that of certain species of *Trematomus* which live beneath the Antarctic ice (Grigg, 1967). The haemoglobins of these species have not been characterised but it is unlikely that they are monomeric. This suggests that the haem-haem interactions which are believed to cause the sigmoid oxygen dissociation curve do not necessarily occur in all fish with tetrameric haemoglobins.

Active fish inhabiting waters of high oxygen content tend to have sigmoid oxygen curves, e.g. *Salmo gairdneri* (Beaumont and Randall, 1969), *Lates albertianus* (Fish, 1956), *Scomber scombrus* (Root, 1931). The sigmoid shape reflects the right shift of the curve combined with the steep middle segment. This implies that at the partial pressure present in the tissues large quantities of oxygen can be unloaded without the tissue requiring to be subjected to a low P_{O_2} to gain access to this oxygen store. Yet the high P_{O_2} of the respired water ensures full oxygenation despite the displacement of the curve to the right.

The shape of the dissociation curve is influenced by temperature. A rise in temperature tends to shift the curve to the right, and to flatten it, presumably by enhancing haem–haem interaction (Dill, Edwards and Florkin, 1932). This must hinder oxygen uptake in small bodies of water warmed by the sun, as the increased difficulty in binding oxygen at the gills will coincide with an increased demand for it by the tissues, and a decreased solubility of oxygen in water at higher temperatures. Black, Kirkpatrick and Tucker (1966) report that in *Salvelinus fontinalis* acclimated to low temperatures, the increase in the oxygen affinity of the haemoglobin which would tend to hinder unloading in the tissues is offset by an increased Bohr effect. The haemoglobins of the Antarctic fish *Trematomus bernacchii* and *T. borchgrevinki* are adapted to function at the very low temperatures -1.4 to $-2°$ C which occur there. The oxygen affinity of their blood is greatly diminished by even small increases of temperature; moreover the oxygen capacity is reduced. At $4.5°$ C the oxygen capacity is only 70% of normal (Grigg, 1967). The Antarctic fish emphasise for us the remarkable range of temperature over which haemoglobin is adapted to function in different species of fish. The characteristic temperature to which each haemoglobin is specifically adapted is presumably a property imparted to the haem by the globin to which it is attached.

B *The oxygen affinity of fish blood*
The oxygen affinity of fish blood expressed as the P_{50}, tends to be related to the environmental oxygen pressures to which the fish is exposed. In table 7 the contrast between the blood of the brook trout, *Salvelinus fontinalis*, which inhabits fast-flowing streams, and *Cyprinus carpio*, which lives in ponds and rivers with a lower oxygen content is evident. Active fish require blood which will unload oxygen in the muscles at a high partial pressure. Fast sustained swimming is not possible if the tissue P_{O_2} must fall to low levels before the haemoglobin unloads its bound oxygen. The P_{50} of the Mako shark, *Isurus oxyrinchus*, may be compared with that of *Scyliorhinus stellaris*, in this regard. The brown bullhead, *Ictalurus nebulosus*, in contrast can live in waters of low oxygen content. Black (1940) reports that its P_{50} is 1.4 mmHg. Such a fish can still load its haemoglobin with oxygen at pressures which render other species, e.g. *Salmo gairdneri*, unable to utilise such oxygen as is in the water. Marvin and Heath (1968) report that *Ictalurus nebulosus* will survive in poorly oxygenated water in which *Salmo gairdneri* die. But *Ictalurus* must endure a very low partial pressure of

TABLE 7. *The oxygen affinity of the bloods of various species of fish, expressed as the* P_{50}

Species	Temperature (°C)	P_{CO_2} mmHg	P_{50} mmHg	Author
Scyliorhinus stellaris	17	1·4	12·0	Piiper and Schumann (1967)
Isurus oxyrinchus	25	1–2	24·0	Lenfant and Johansen (1969)
Salmo gairdneri	15	0–1	43·0	Beaumont and Randall (1969)
Salmo salar	15	0·0	10·0	Black, Tucker and Kirkpatrick (1966)
Salvelinus fontinalis	15	0·0	12·0	Black, Kirkpatrick and Tucker (1966)
Ictalurus nebulosus	15	0–1	1·4	Black (1940)
Trematomus borchgrevinki	−1·5	0–1	21·0	Grigg (1967)
Cyprinus carpio	15	1–2	5·0	} Black (1940)
Catostomus commersoni	15	1–2	12·0	

oxygen in its tissues if unloading is to occur. The embryo of the oviparous *Raja binoculata* has a haemoglobin with a higher oxygen affinity than the foetus and adult (Manwell, 1958*b*). In the ovoviviparous *Squalus suckleyi* where gestation takes 22–23 months there is a definite foetal–maternal shift of the oxygen dissociation curve to the right (Manwell, 1958*c*). In addition there are haem–haem interactions in the foetal haemoglobin ($n = 1·23$) which are absent in the adult ($n = 1$).

c *The Bohr effect*

An increased concentration of hydrogen ions causes configurational changes in the haemoglobin molecule that inhibit the binding of oxygen. If the points of contact between the sub-units are ionisable groups with pK values within the pH range in which oxygen is loaded and unloaded, the configurational changes and equilibrium values will become pH dependent (Riggs, 1965). Both the monomeric haemoglobin of *Petromyzon marinus* (Wald and Riggs, 1951) and the tetrameric haemoglobin of teleost fish exhibit a Bohr effect suggesting that more than one type of sub-unit interaction may be involved.

Fish that inhabit waters rich in oxygen commonly have a

haemoglobin with a marked Bohr effect; e.g. the sucker, *Catostomus commersoni*, has a P_{50} of 12 mmHg at a P_{CO_2} of 1-2 mmHg. If the P_{CO_2} is increased to 18–20 mmHg the P_{50} rises to 66 mmHg (Black, 1940). The fish *Lates albertianus*, endemic to Lake Albert, is amongst the largest of freshwater fish and exhibits a well-developed Bohr effect. Its P_{50} is 17 mmHg at 0 mmHg of CO_2, but increases to 34 mmHg at 25 mmHg of CO_2 (Fish, 1956). The Bohr effect confers a distinct physiological advantage to fish living in water of high oxygen and low carbon dioxide content. Black (1940) points out that if the blood of *Catostomus commersoni* were to encounter a P_{CO_2} of 20 mmHg as it passed through the tissues it would discharge 85% of its bound oxygen without any decrease in P_{O_2}. Fish which inhabit acidic waters of low oxygen content cannot avail themselves of the advantages that the Bohr effect can offer as it would prevent the loading of oxygen at the gills. Such fish may possess a haemoglobin remarkably insensitive to hydrogen ion concentration. That of *Mormyrus kannume*, a bottom-living fish of Nile streams, is fully saturated at a P_{O_2} of 5 mmHg and this figure is unchanged if the P_{CO_2} is increased to 20 mmHg (Fish, 1956). *Hoplostomum littorale*, a fish of Paraguayan swamps, has a haemoglobin that is virtually unaffected by a P_{CO_2} of 25 mmHg (Willmer, 1934).

Of the two haemoglobins in the blood of *Oncorhynchus keta*, the S component has a high oxygen affinity that hardly changes between pH 5·7–8·4. The F component has a low oxygen affinity and exhibits a large Bohr effect; it would not be more than 20% saturated at a pH of 6·8 and atmospheric pressure of oxygen (Hashimoto, Yamaguchi and Matsuura, 1960).

It is known that the oxygen affinity of mammalian haemoglobins can be depressed by molecular CO_2 independently of any changes in pH that it may exert. Carbon dioxide can effect oxygen equilibria by forming carbamino compounds with haemoglobin, at constant pH. We know little of this aspect of the Bohr effect in fish blood.

D *The Root effect*

In certain species of fish, Root (1931) and Root and Irving (1941) observed that an increased partial pressure of carbon dioxide not only decreased the affinity of the haemoglobin for oxygen, but rendered a proportion of it incapable of binding oxygen at any pressure. It, in effect, decreased the oxygen capacity of the blood. Scholander and Van Dam (1954) showed that in some deep-sea fish, e.g. *Alphestes*, *Epinephelus*, a pH of 6·7–5·6 prevented the blood from becoming more than 30–80% saturated even at 140 atmospheres pressure of oxygen. The Root effect may be regarded as an extreme case of the Bohr effect; it is much less marked in haemolysed blood and requires a greater change of pH and P_{CO_2} than in the intact red cell. Manwell (1964) suggests that the Root effect is a pH–dependent negative haem–haem interaction. When the hydrogen ion concentration is increased, the binding of oxygen to one haem group brings about configurational changes which prevent oxygen binding to an adjacent group.

The off-loading of oxygen caused by the Root effect is known as the Root-off shift. It has been shown to have a shorter half time than the Root-on shift. The importance of these different rate constants in the deposition of oxygen in the swim-bladder has already been discussed.

The transport of carbon dioxide

Rahn (1966) has emphasised that there are certain physical properties of carbon dioxide compared with oxygen, which go far towards explaining its low partial pressure in the tissues of fish. Carbon dioxide is 28 times more soluble in water than oxygen at 20° C. Fish, like other vertebrates, are involved in exchanging the one gas for the other. If we assume a respiratory quotient of 1, the consumption by a fish of an amount of oxygen that would lower its partial pressure in water by 28 mmHg would return to the water only sufficient CO_2 to raise its partial pressure by 1 mmHg. This great difference in the solubility of the two gases guarantees a low partial pressure of carbon dioxide in the tissues (Hughes, 1966c). The low solubility of oxygen in water enforces a high rate of ventilation which inevitably washes out the carbon dioxide. Two other features help to maintain low carbon dioxide pressures in fish. There is very little carbon dioxide in most natural waters and the presence of bicarbonates tends to buffer much of what is added by living organisms. Thus the gradient of carbon dioxide partial pressure between the blood and the water is almost the maximum possible. The respiratory system of fish serves to maintain a unidirectional flow of water across the respiratory surfaces. There is no quantity of water comparable to the alveolar air of a mammal in which carbon dioxide can accumulate. This absence of an anatomical dead space, may be in part offset by the presence of a physiological dead space caused by the incomplete circulation of blood and water past the respiratory surfaces (Hughes, 1964). Nevertheless, these three features of the aquatic respiration of fish combine to maintain a partial pressure of carbon dioxide in the plasma and tissues which seldom exceeds 10 mmHg. Hughes (1964) has discussed the effectiveness of fish gills in respiration. The term 'effectiveness' refers to the ability of gills to transfer oxygen to and remove carbon dioxide from the blood. Viewed in this way, the gills are very effective in removing carbon dioxide, as can be seen when the partial pressure of the gas in blood drawn from the ventral and dorsal aorta are compared. Typical values are *Squalus suckleyi* 2·5 and 1·7 mmHg (Lenfant and Johansen, 1966), *Scyliorhinus stellaris* 3·3±1·3, 1·9±0·5 mmHg (Piiper and Schumann, 1967), *Salmo gairdneri* 5·7±1·5, 2·3±1·1 mmHg (Stevens and Randall, 1967*b*).

Carbon dioxide is transported in the blood as bicarbonates and perhaps, as carbamino compounds. The carbon dioxide combining power of a blood represents the amount of carbon dioxide in the blood at a particular partial pressure. Fish living in poorly oxygenated environments tend to have blood with a greater carbon dioxide combining power than those from well aerated water, e.g. carp 21 vols% at a P_{CO_2} of 3·8 mmHg, *Prionotus carolinus* 8·5 vols% at a P_{CO_2} of 3·5 mmHg (Lenfant, Johansen and Grigg, 1966/67).

If the total carbon dioxide content of blood is plotted against its P_{CO_2}, a carbon dioxide absorption curve is obtained. Such curves exhibit a steeply-rising initial segment within which increasing carbon dioxide content effects little change in P_{CO_2}, because it is taken up as bicarbonate. This segment spans the partial pressures generated within the tissues under normal conditions. At partial pressures above 10–20 mmHg the slope of the curve rises less steeply.

Part of the buffering power of the blood lies in the plasma, and part within the red cells. Oxygenated haemoglobin is a stronger acid than deoxygenated haemoglobin so that carbon dioxide is displaced from the blood as oxygen is loaded on in

73

the gills. This effect, which is the equivalent for carbon dioxide that the Bohr effect is for oxygen, is known as the Haldane effect.

Fish bloods differ greatly in the magnitude of the Haldane effect. It is absent in *Squalus suckleyi* (Lenfant and Johansen, 1966), and in *Raja oscillata* (Dill *et al.*, 1932). Salmonid fish show a marked Haldane effect, greater in *Salvelinus fontinalis* (Black, Kirkpatrick and Tucker, 1966) than in *Salmo gairdneri*. Lenfant *et al.* (1966/67) question whether the Haldane effect is of any adaptive value and suggest that variations in it reflect fundamental differences in haemoglobin structure. There are also marked differences in the contribution made to the buffering power of the blood, by the plasma and the red cells. In *Squalus suckleyi* (Lenfant and Johansen, 1966), and in *Raja oscillata* (Dill *et al.*, 1932) the major part of the buffering power is contributed by the plasma proteins. Skate plasma proteins have, per gram, about twice the intrinsic buffering power of human plasma proteins. In *Neoceratodus forsteri* most of the buffering power resides in the red cells (Lenfant *et al.*, 1966/67).

Fish that are primarily air breathers, such as *Electrophorus electricus*, and the lung fish *Protopterus* and *Lepidosiren*, are less able to dispose of their carbon dioxide than are gill breathers. The high solubility of carbon dioxide in water favours, as we have seen, the aquatic elimination of this gas. Obligate air-breathing fish thus tend to have blood with a high P_{CO_2} and a high buffering power. The P_{CO_2} of the arterial blood of *Electrophorus* is 27·7 mmHg (Johansen *et al.*, 1968*b*). The buffering power of the blood of the obligate air-breathing *Lepidosiren* is considerably greater than that of *Neoceratodus*, which still has fully functional gills (Johansen and Lenfant, 1967). The superiority of gills as organs of carbon dioxide elimination, and the greater abundance of oxygen in air, as compared with water, has tended to produce a bimodal gas exchange in such fish (Hughes, 1966*c*). The principal role of the air-breathing organ becomes oxygen absorption; the gills may still retain a crucial role in the elimination of carbon dioxide (Johansen, 1966; Johansen and Lenfant, 1967, 1969; Johansen *et al.*, 1967). In *Protopterus*, Jesse, Shub and Fishman (1967) report that the stimulus of 8% oxygen increases lung ventilation by three times but gill ventilation by only one-half, and they suggest that separate receptors are involved. The various species of air-breathing fish are of great interest in exhibiting a spectrum of the stages through which the terrestrial vertebrates must have passed, and during which the blood must have acquired its ability to accommodate and buffer increasing quantities of carbon dioxide.

The regulation of oxygen binding power

It was widely believed that the oxygen dissociation curve of a haemoglobin, though affected by pH and temperature, was otherwise a relatively stable feature of the blood of an organism. During the last decade it has become apparent that numerous ligands can change the affinity of the haemoglobin for oxygen. In this context the term ligand refers to any substance that binds to haemoglobin and thus alters its properties. The production of ligands may constitute an intrinsic adaptive mechanism in response to various stimuli such as hypoxia. Benesch and Benesch (1967) reported that if human haemoglobin is dialysed free of electrolytes its oxygen dissociation curve exhibits no significant haem–haem interactions, and the oxygen affinity is very high. The addition of Na_2SO_4 or $NaCl$ effects no marked change, but organic phosphate, and in particular 2,3-diphosphoglycerate (2,3-DPG), restores haem–haem interactions to their original level. The oxygen affinity of human haemoglobin deprived of 2,3-DPG is equal to that of myoglobin (Benesch and Benesch, 1969).

This particular organic phosphate is abundant in mammalian red cells. The 2,3-DPG molecule binds only to deoxygenated haemoglobins. The 2,3-DPG and oxygen are mutually exclusive. The combination of 2,3-DPG with haemoglobin occurs in the dyad axis in the central cavity where arginines and lysines are present to neutralise its negative charge, and is with the beta rather than the alpha chains (Perutz et al., 1968). It is the widening of the gap between the beta chains during deoxygenation that facilitates the binding of 2,3-DPG.

The stimulus of hypoxia leads, in mammals, to an increase in the level of red cell 2,3-DPG whether it is caused by living at high altitudes (Lenfant et al., 1968), or by some form of anaemia (Lenfant et al., 1970). The response is rapid, occurring within 48 hr, and its speed of onset falls into an intermediate position

75

between the cardiovascular response to hypoxia, which occurs in a few minutes, and the increase in haemopoiesis which occurs after some weeks. It is an adaptive response because the decreased affinity for oxygen, or shift to the right of the dissociation curve enhances the unloading of oxygen in the tissues. The red cell, it appears, possesses an allosteric system which regulates the oxygen affinity of its contained haemoglobin (Benesch and Benesch, 1969). Hypoxia leads to the binding of 2,3-DPG and the depletion of intracellular soluble organic phosphate. This stimulates glycolysis and the increased production of 2,3-DPG; accumulation of 2,3-DPG in turn inhibits the early stages of the enzyme sequence leading to its own production. The mechanism can automatically control the balance between oxygenated and deoxygenated haemoglobin.

It appears that different vertebrate classes may utilise different soluble organic phosphates to this end. In birds 2,3-DPG is replaced by inositol hexophosphate (IHP) which appears to operate in a similar way (Benesch and Benesch, 1969). Both ATP (adenosine triphosphate) and ADP (adenosine diphosphate) are active in this regard and Akerblom *et al.* (1968) found a linear relationship between the sum of the molar concentrations of 2,3-DPG plus ATP and the P_{60} of stored human blood.

Soluble organic phosphates are certainly present in fish red cells and Lenfant (1969) reports the following data: *Amia calva* 3·5 mM 2,3-DPG, 4·9 mM ATP; *Acipenser* 0·5 mM 2,3-DPG, 1·6 mM ATP per litre of red cells. Rapoport and Guest (1941) could find no 2,3-DPG in the blood of *Ictalurus* and *Micropterus* but ATP and an unidentified organic phosphate were present. Whilst the identity of the regulatory ligands in fish remains uncertain, it is likely that such regulation exists. Grigg (1969) reports that the oxygen affinity of the blood of *Ictalurus nebulosus* is increased by acclimation to high temperature. The P_{50} of the blood of warm acclimated fish is 8·2 mmHg; that of cold acclimated fish is 13 mmHg. Houston and De Wilde (1968*b*) postulate some adjustment of the oxygen transport mechanism which can amplify oxygen flow through the system, in their study of thermally acclimated carp.

It remains an exciting hypothesis that, for example, the acclimation of *Salvelinus fontinalis* to hypoxia, described by Shepherd (1955), in part depends upon this type of regulatory mechanism. Indeed, we might speculate further that in cold-blooded vertebrates, in which the internal environment is regulated less precisely, the changes in temperature and ionic environment to which the haemoglobin is subjected, make such a mechanism more imperative.

CHAPTER 7

The Autonomic Nervous System and the Vascular Receptors

Our knowledge of the anatomy of the autonomic nervous system of fish still rests largely on two major studies by Young, on *Uranoscopus scaber* (1931) and on *Scyliorhinus canicula* and other elasmobranch fish (1933). West (1955) has discussed the relationship which the scattered groups of chromaffin cells of elasmobranch fish bear to the adrenal medullary tissue of higher vertebrates. Nicol described the autonomic nervous system of the chimaeroid fish *Hydrolagus colliei* (1950) and (1952) reviewed the autonomic nervous system of fish in relation to that of other vertebrates. Burnstock (1969) has discussed the evolution of autonomic nervous control of the visceral and cardiovascular systems of lower vertebrates. It is necessary to consider briefly some of the more important findings of these and other studies because the level of development of the fish autonomic nervous system defines certain limits within which cardiovascular regulation can occur.

The innervation of the heart

Within the Cyclostomata, the hagfish *Myxine* and *Eptatretus* have an aneural heart which receives no innervation from the vagus (Greene, 1902; Jensen, 1961; Falck *et al.*, 1966). Hirsch, Jellinek and Cooper (1964) report the presence of nerve fibres and ganglia in the myocardium of *Eptatretus*, and contest the conclusions of earlier studies, but the balance of evidence seems at present to be that there is no cardiac vagal supply in this genus. The hearts of the hagfish do not respond to vagal stimulation and are insensitive to applied acetylcholine. The lampreys in contrast have a heart innervated by the vagus. It exerts its cardioregulatory influence in a manner opposite to that of all higher vertebrates. Stimulating this nerve or

78

applying acetylcholine to the heart causes cardioacceleration; the response is insensitive to atropine but is blocked by curare (Augustinsson *et al.*, 1956).

Elasmobranch fish have a heart innervated by two nerves; one is derived from the last branchial branch of the vagus, and the other from its visceral branch (Young, 1933). Both are inhibitory to the heart; their action and that of applied acetylcholine are blocked by atropine. Teleost fish have a single cardioinhibitory nerve which arises from several roots from the visceral branch of the vagus (Young, 1931).

It has been widely maintained that the heart of fish lacks sympathetic innervation. This conclusion, originally put forward by Young (1931, 1933) was supported by the electron microscope studies of Conteaux and Laurent (1958). However, there is a possibility, pointed out by Young (1931, 1933) that sympathetic fibres enter the vagus in *Scyliorhinus* through branches from the first sympathetic ganglion, or, in teleost fish, through rami communicantes from more cranial outflows. Gannon and Burnstock (1969) have succeeded, using fluorescent histochemical techniques in demonstrating catecholamine-containing nerve fibres in the sinus venosus, atrium and ventricle of *Salmo trutta* and *S. irideus*. *Salmo irideus* is the name commonly given to the freshwater phase of *S. gairdneri*. Stimulating the vagus of these species when the cholinergic fibres are blocked with atropine, causes cardioacceleration. This response is blocked by beta blocking agents. Although these results suggest that the adrenergic fibres in the teleost heart run in the vagus there is evidence that those destined for the ventricle run in with the coronary arteries. Nerve trunks running in the perivascular plexus around the coronary arteries both penetrate the media to the vascular smooth muscle and give off branches to the myocardium. Shelton and Randall (1962) report that cutting the vagus of an anaesthetised tench causes the heart to slow and suggest that cardioacceleratory fibres may be present. The balance of evidence would now seem to suggest that there is a direct sympathetic innervation of the fish heart.

The innervation of peripheral blood vessels

The sympathetic system of Cyclostomata (Nicol, 1952) consists only of scattered nerve cells or ganglia lying along the cardinal

79

veins; they are not joined by longitudinal connectives. From these neurons some fibres pass to blood vessels but their function has not yet been studied.

In elasmobranch fish (Young, 1933) the sympathetic ganglia form an irregular longitudinal series between the levels of the pectoral and pelvic fins. They are linked by irregular longitudinal connectives but no well-defined chain is present. Myelinated fibres pass from the spinal nerves through white rami to the sympathetic ganglia; unmyelinated post-ganglionic fibres pass from them and run along the blood vessels, some of which they innervate. There are no recurrent grey rami returning fibres to be distributed with the spinal nerves. The blood vessels of the head and post-pelvic trunk must receive their vasomotor supply, if they have one, as extensions of the intramural net of fine nerve fibres known to invest certain blood vessels such as the caudal artery.

Teleost fish have a more developed sympathetic system with both longitudinal and transverse connectives. Teleosts are unique amongst the vertebrates in that the sympathetic chain extends forwards into the head and bears ganglia in connection with the cranial nerves (Young, 1931). From the spinal sympathetic ganglia post-ganglionic fibres pass to the blood vessels of the viscera; others return to the spinal nerves as grey rami. It is known that they are distributed within the spinal nerves to the chromatophores. Kirby and Burnstock (1969) report that isolated strips of the ventral aorta of eel and trout contract in response to electrical stimulation and conclude, from pharmacological studies, that these contractions are in part mediated by post-ganglionic adrenergic nerve fibres. Read and Burnstock (1968) have identified adrenergic nerve fibres in the gut vessels of eel and trout by fluorescent histochemical techniques. Gannon and Burnstock (1969) report that fine nerve fibres form a perivascular plexus around the coronary arteries of *Salmo trutta* and *S. irideus*. Bulger and Trump (1969) have studied the ultrastructure of arterioles in the kidney of salt and freshwater teleost fish and report the presence of nerves and associated ganglion cells there. Mott (1950) reports that the smooth muscle in the bulbus arteriosus of the eel can be seen by radiography to contract spasmodically. These observations suggest that there is a peripheral vasomotor nerve supply in fish, and the presence of grey rami in teleost fish suggests that this may be more extensive in such fish than in the Elasmobranchii.

Chromaffin tissue

In elasmobranch fish a pair of 'axillary bodies' lie above the posterior cardinal sinus just behind the sinus venosus. The two are not symmetrical in size or position; each consists of a mass of chromaffin tissue coalesced with a number of fused sympathetic ganglia. From the ganglia post-ganglionic fibres run to the chromaffin cells (Young, 1933). These cells are known to be rich in catecholamines of which noradrenaline constitutes at least 50% (Shepherd, West and Erspamer, 1953). Their strategic location in the path of the incoming blood suggests that catecholamines can be brought to the myocardium with a minimum of delay by liberation into the blood flowing into the heart. Behind the axillary bodies lie similar but smaller aggregations of sympathetic neurons and chromaffin cells; the more posterior of these form a segmental series embedded within the kidney tissue.

The cortical tissue in elasmobranch fish forms a median spindle-shaped mass, the interrenal body, that is not closely associated with the chromaffin tissue.

In teleost fish chromaffin cells are distributed primarily around the course of the posterior cardinal veins. In some families, e.g. Salmonidae and Catostomidae they are quite separate from the interrenal tissue; in others, e.g. the Cichlidae and Labridae, the two tissues are intermingled (Nandi, 1962).

Receptors of the cardiovascular system

The pseudobranch of teleost fish has been studied by Laurent and Dunel (1964, 1966). It is derived from the gill-like structure present in elasmobranch fish and like it receives its blood supply from the afferent pseudobranchial artery. It is thus perfused with arterialised blood, derived from the efferent loop of the first gill.

Thick myelinated nerve endings occur in the walls of the main vessels afferent to the pseudobranchial leaflets of the tench and the black bass; they degenerate if the pseudobranchial nerve, a branch of the glossopharyngeal nerve, is cut. They are therefore primary afferent endings. Laurent (1967) believes these to be baroreceptor endings and reports that nerve discharges can be recorded in isolated fibres from the glossopharyngeal nerve; these discharges increase in frequency as the perfusion pressure is increased. Randall and Jones (1969) report similar findings from rainbow trout.

Laurent and Dunel (1966) report that there are three layers of fine nerve endings in the pseudobranch. The outer, or first layer, occurs in the efferent arteries; the second plexus is in the axis of each plumule; the third, or interlamellar plexus, occurs between the layers of pseudobranchial cells. These fine nerve endings are the terminations of fibres arising from small bipolar nerve cells in the pseudobranchial tissue, the axons of which synapse with large multipolar neurons. The bipolar cells thus do not degenerate when the pseudobranchial nerve is cut. Laurent and Dunel (1966) believe these fine endings to be chemosensitive. Those of the interlamellar plexus come into association with the pseudobranchial cells which Copeland and Dalton (1959) have shown to have tubules of endoplasmic reticulum opening to the surface of the plasma membrane and thus into the vascular spaces. Microelectrode recordings from the pseudobranchial cells (Laurent, 1967) of isolated pseudobranchs show that these cells are depolarised if the pseudobranch is perfused with a fluid of low P_{O_2} or raised P_{CO_2} This depolarisation is accompanied by increased firing in afferent fibres in the pseudobranchial nerve. Acetazolamide hyperpolarises the pseudobranchial cells and inhibits the stimulating effect of carbon dioxide. These studies suggest that the pseudobranch is a specialised pressure and chemosensitive structure. Its location in the pathway that carries blood to the head recalls that of the carotid sinus and body in mammals. However, many families of teleost fish lack the pseudobranch altogether and Satchell (1961) demonstrated that ablating the gill-like pseudobranch of *Squalus acanthias* in no way reduced the reflex bradycardia evoked by hypoxia.

There have been indications that baro- and chemoreceptors may be more widely distributed in other species. Boyd (1936) reported a rich innervation of the junctions of the efferent branchial arteries with their arterioles in *Mustelus*, and suggested that the fine nerve endings may be baroreceptors. De Kock (1963) described a richly innervated region where the first epibranchial artery enters the dorsal aorta of the salmon and trout. Fine nerve fibres and ganglion cells derived from the glossopharyngeal nerve are associated with a region of the vessel wall in which the muscular coat is thin, and the endothelial cells are replaced by large cells with lobulated nuclei. The second, third and fourth epibranchial arteries have no such structure. The first epibranchial artery of a teleost is homologous with the carotid artery of a mammal and the structures described by De Kock may be the forerunner of the carotid body of higher vertebrates. No functional analysis of these endings has been made.

Teleost fish possess numerous sensory endbuds (De Kock, 1963) which form a linear series along the inner surface of the gill arches and spread onto the pharyngeal wall. Each is a bulb-shaped end organ situated within a hillock of epidermis; within it sensory cells with elongated nuclei are surrounded by supporting cells. Each sensory cell sends a hair-like projection through an epidermal pore. These receptors may be taste buds and have only a gustatory function; De Kock (1963) suggests that they may also be chemoreceptors that monitor the inspired water. Shepherd (1955) reports that *Salvelinus fontinalis* become violently active when first introduced into water of low oxygen content. Johansen and Lenfant (1967) conclude that the chemoreceptors of *Lepidosiren paradoxa* face into the water.

Receptors may also be located on the venous side of the circulatory system. In *Scyliorhinus canicula* and certain other elasmobranch fish there is a plexus of fine nerve fibres behind the heart, extending back to the anterior end of the kidney. It is derived from fibres passing out from the anterior spinal nerves and intermingling with fibres from the vagus. Within this post-branchial plexus (Young, 1933) are afferent fibres derived from large encapsulated sense organs, each of which consists of whorls of fine fibres bearing swellings at intervals. They are located in the tissue above the posterior cardinal sinus and are thus well positioned to monitor central venous pressure; no functional studies have been made of them.

CHAPTER 8

The Control of the Heart

The organism requires to regulate the output of the heart so that it can, within limits, be matched to the circulatory demands of the tissues. The two factors that determine the cardiac output are the stroke volume and the heart rate. These may alter because the myocardium possesses an intrinsic response to the pressure of the returning blood. They may also be altered by regulatory influences imposed by changes in the level of circulating catecholamines or by the direct influence of the cardioregulatory nerves.

The intrinsic responses of the myocardium

In the isolated mammalian heart stroke volume is related to right atrial filling pressure; this relationship is known as Starling's law of the heart. Increased right atrial filling pressure, and thus increased stretch of the myocardium, causes an increased stroke volume with no change in heart rate.

In the aneural heart of the hagfish *Eptatretus*, Jensen (1961, 1965) has shown that increased perfusion pressure increases the heart rate. The isolated ventricle can increase its rate by 50–150%; the response to distention occurs almost immediately and continues until the excess of fluid is removed. In the isolated heart of *Salmo gairdneri*, Bennion and Randall (1970) have shown a similar dependence of heart rate on perfusion pressure. The heart rate increases from 35–58/min as the perfusion pressure is increased from 2–12 mmHg. The phenomenon is well known in other vertebrates and has been demonstrated in the isolated hearts of tortoise, frog and rabbit (Keatinge, 1959), dog and rabbit (Blinks, 1956), and frog (Pathak, 1958). It represents an intrinsic response of the heart to increased filling pressure quite distinct from that embodied in Starling's law. Jensen (1961) has suggested that the pacemaker cells of the heart are sensitive to stretch and depolarise more rapidly when extended.

That the myocardium of the fish heart can in addition, respond to increased filling pressure by an increase in stroke volume has been demonstrated in *Eptatretus* by Chapman, Jensen and Wildenthal (1963) and in *Salmo gairdneri* by Bennion and Randall (1970). In *Eptatretus* a stepwise increase in filling pressure evokes an increase in the force of contraction up to a level at which the heart stops altogether. The curve relating pressure to tension resembles a Starling curve with the descending limb omitted. In this species the myocardium contains a potent cardiostimulatory agent, an amine or an amide, called 'eptatretin' which is also a cardiac stimulant to the heart of dog and frog; it has been suggested that it plays some part in the intrinsic responses of the heart of *Eptatretus* (Jensen, 1963). In *Salmo gairdneri* elevating the filling pressure from 2–10 mmHg increased the stroke volume from 0·03–0·058 ml/beat (Bennion and Randall, 1970).

These intrinsic responses of rate and tension of the fish myocardium provide a basis for cardioregulation. An increase in venous return caused by the pumping action of muscles on veins, and the dilation of small muscle vessels resulting from metabolic activity could initiate an increase in cardiac output.

Extrinsic cardioregulation

It is known that during exercise there is a rise in the level of circulating catecholamines in the blood of the rainbow trout (Nakano and Tomlinson, 1967). Bennion and Randall (1970) have shown that in this species the isolated heart shows an enhanced response to an increase of filling pressure as the level of catecholamines in the perfusate is increased. The increase of stroke volume from 0·03 ml/beat to 0·058 ml/beat, effected by elevating the perfusion pressure from 2–10 mmHg, could be augmented to 0·08 ml/beat when the concentration of adrenaline was raised from 0·01 μg/ml to 0·1 μg/ml. These levels correspond to the actual blood catecholamine levels in resting and disturbed trout respectively (Nankano and Tomlinson, 1967). Successive increases in the level of adrenaline in the perfusate permitted a family of curves relating pressures to stroke volume, to be plotted one above the other. It would thus seem that there is an indirect control of cardiac output in this fish effected by the central nervous system through regulating the liberation of catecholamines into the circulating blood. However, this may not be true of the Cyclostomata. Catecholamines are abundant in the heart muscle of *Myxine* (Östlund, 1954) but these agents appear to have no influence on the normal activity of the Cyclostome heart (Chapman *et al.*, 1963; Augustinsson *et al.*, 1956).

The regulation of blood pressure in mammals through the operation of the carotid sinus reflex in which increased baroreceptor activity evokes a fall in blood pressure by the reflex adjustment of cardiac output and peripheral resistance, has prompted many workers to search for similar mechanisms in fish. Irving, Solandt and Solandt (1935) reported that in *Mustelus* increasing the blood pressure in the gill vessels slowed the heart and that bursts of action potentials could be observed in afferent nerve fibres from the branchial area; these discharges occurred in time with the heart beat. Increasing the blood pressure by injecting adrenaline caused the discharges to increase; lowering it by bleeding the fish caused them to disappear. Brief elevations of blood pressure in the branchial vessels caused cardiac slowing in *Squalus acanthias* (Lutz and Wyman, 1932) and in the eel (Mott, 1951). Small injections of adrenaline into the coho salmon (*Oncorhynchus kisutch*) bring about an increase of blood pressure and a fall in heart rate; after blocking the cholinergic inhibitory fibres to the heart with atropine, similar injections of adrenaline caused a larger increase of blood pressure and an increase in heart rate (Randall and Stevens, 1967). This could be explained if some of the resting vagal tone had its origin in a baroreceptor reflex.

Despite this evidence some doubt still remains about the existence of a reflex regulation of blood pressure in fish comparable to that of mammals. More than 30 years have elapsed since Irving *et al.* (1935) published their observations; in that time, no one has succeeded in demonstrating that sustained changes in blood pressure evoke sustained alterations of heart rate. Indeed, Sudak and Wilber (1960) concluded from a study of the response to haemorrhage and reinfusion of blood into *Mustelus canis* and *Squalus acanthias* that their circulatory systems behave as simple elastic reservoirs and that there is no evidence, in these fish, of any pressure compensatory mechanism.

86

That there is a continuous tonic vagal inhibition of the heart is well recognised, and species vary as to the intensity of this. The sucker, *Catostomus macrocheilus*, has a higher vagal tone than the rainbow trout, *Salmo gairdneri* (Stevens and Randall, 1967*a*). But this tonic vagal activity is not necessarily the output of a baroreceptor reflex. It may be modulated in the service of any other reflex in which it is necessary to regulate the beat of the heart.

Cardiorespiratory synchrony

Unlike most mammals the hearts of many fish beat at a slower rate than that of respiration, and Lyon (1926) noted that in certain elasmobranch fish, heart and respiration were co-ordinated. Satchell (1960) analysed this synchrony in *Squalus acanthias* and *Mustelus antarcticus* and showed that it was re-flexly mediated. The heart tends to beat at every second, third or fourth respiration and the P wave of the electrocardio-gram occurs as the mouth opens. Some, as yet unidentified, receptor is stimulated, either by the moving stream of respira-tory water, or by the movements of respiration. The action potentials so generated, pass up the branchial nerves to the medulla where they reflexly generate activity in the cardiac vagus and inhibit the heart. Shelton and Randall (1962) have reported cardiorespiratory synchrony in teleost fish. Curarisa-tion abolishes synchrony, suggesting that it is effected by a reflex peripheral loop rather than some central linkage of vagal and respiratory neurons within the medulla.

In many fish cardiorespiratory synchrony may be absent or is a transient phenomenon. In *Callionymus lyra* the heart beat is more frequent than respiration; ratios as high as five beats per respiratory cycle have been reported (Hughes and Umezawa, 1968). In this fish, and in *Salmo gairdneri* (Hughes and Roberts, 1970) varying degrees of coupling occur and there is much individual variation.

The nature of the receptor responsible for cardiorespiratory synchrony is unknown but there is some evidence that chemo-receptors monitoring oxygen pressure of either the inspired water or the arterialised blood may be involved. In *Salmo gairdneri* Randall and Smith (1967) report that cardiorespira-tory synchrony is not evident at normal oxygen pressure, but appears when it is reduced to 80–100 mmHg. Its appearance

87

coincides with the onset of hypoxic bradycardia. If the pharynx is artificially ventilated with a hand pump a new rhythm can be imposed upon the heart. In the tench (Randall, 1966) bursts of action potentials can be recorded in the cardiac vagus which are synchronous with respiration; hypoxia causes these trains to be lengthened and firing to become more intense. Each train of action potentials is initiated as the mouth opens; during the inspiratory phase of respiration the buccal pump cannot force water over the gills. As the mouth closes the irrigation of the gills coincides with the end of the train and permits the heart to beat. These results suggest that the bradycardia of hypoxia and cardiorespiratory synchrony are mediated by the same reflex pathway and that the heart beat comes to be entrained by respiration because the cyclic perfusion of the gills effects a cyclic reduction of the cardiac inhibition that hypoxia generates. The question, however, must remain unsettled until further work has been done on a greater variety of fish.

When the coupling of the heart beat to respiration is reduced various patterns of cardiac arrhythmia are generated. These can be duplicated by a computer if the assumption is made that a saw-toothed wave form, representing the activity of the cardiac pacemaker, is cyclically modulated by a sinusoidal wave form representing the activity generated by respiration and relayed to the heart by the vagus. The most common arrhythmia is one in which there is a regular occurrence of cycles that are either markedly longer or shorter than the other cycles in the trace. Short cycles are followed by cycles of increasing length; long cycles are followed by cycles of decreasing length (Satchell, 1968). These arrhythmias bear a resemblance to the normal sinus arrhythmia in the mammal heart. Because the heart rate in fish is usually less than the respiratory rate, cyclic activity in the cardiac vagus can trigger particular beats; in mammals, where the respiratory cycle is as long as three or four cardiac cycles, the modulation is seen as an overall increase and decrease of heart rate.

Cardiorespiratory synchrony is a striking feature of many fish electrocardiograms and it is of interest to speculate what circulatory advantages, if any, accrue from it. Shelton and Randall (1962) suggested that it is primarily characteristic of the resting fish; the cyclic activity of the neuronal circuits in the medulla which generate the output of the vagal and respiratory centres, may become entrained when disruptive excitations from extroreceptors are minimal. Satchell (1960) suggested that the exchange of respiratory gases in the gill lamellae between pulsatile flows of blood and water will be most effective if the period of rapid flow of blood into the gills coincides with the period of rapid flow of water across the lamellae. Blood flow would be maximal in the most effective part of the respiratory cycle. In a fish with a 1:2, 1:3 pattern of cardiorespiratory synchrony some respirations will ventilate the gills at a time when the flow of blood through them is reduced. Hughes (1964) has emphasised that synchrony ensures that oxygen is available when blood passes through the gills; it does not necessarily ensure that each ventilation of the gills is accompanied by a corresponding flow of blood. The homeostasis that synchrony achieves is directed to the gas concentrations of the efferent blood rather than the expired water. Since this suggestion was advanced it has been demonstrated that the flow of blood through the gills is indeed pulsatile (Randall *et al.*, 1969). However, the velocity profile of the flow of water between the lamellae is still unknown and the question remains an open one.

CHAPTER 9

The Response to Hypoxia

The small quantity of oxygen contained within a given mass of water compared with the same mass of air clearly imposes a limit on the range of haemorespiratory homeostasis in fish. Nevertheless, there are numerous observations to show that when the oxygen content of the water inspired by a fish is progressively lowered, there is often an initial range of oxygen pressures within which oxygen consumption either remains unchanged or increases. Further reduction below a certain critical value causes a fall in oxygen uptake; within this lower range oxygen consumption is dependent upon oxygen concentration. Some species exhibit this type of response even at partial pressures of oxygen close to atmospheric. In *Scyliorhinus stellaris* (Baumgarten and Piiper, 1969) the oxygen consumption begins to fall as soon as the P_{O_2} of the inspired water falls below 130 mmHg. Fish that respond in this way are said to be oxygen dependent. In contrast, *Salmo gairdneri* (Holeton and Randall, 1967b) responds to lowered P_{O_2} by increasing its ventilation; oxygen consumption rises initially in response to the increased ventilatory work and consumption does not fall below the initial level until the P_{O_2} of the inspired water has decreased to 40 mmHg. Such fish are referred to as oxygen independent. Their consumption is, over a considerable range, independent of the oxygen content of the inspired water.

Salmo gairdneri can increase its ventilation by a factor of 13. Saunders (1962) reports that the stimulus of hypoxia will increase the ventilation of the carp, *Cyprinus carpio*, by 31, of *Ictalurus nebulosus* by 70, and of *Catostomus commersoni* by 82. Under these conditions a 250 g fish will pump its own weight of water over its gills every two seconds. Dependent species do not show so vigorous a response. *Scyliorhinus stellaris* will increase its ventilation by only 25% when the P_{O_2} of the inspired water is lowered from 140 to 50 mmHg (Baumgarten and Piiper, 1969).

Species which are independent at rest may become dependent when their oxygen demands are increased by exercise. *Ictalurus nebulosus* shows little change in its oxygen consumption at rest until the partial pressure falls below 50 mmHg; when active, consumption starts to fall even at 160 mmHg (Basu, 1959). Similarly, an increase of temperature may cause independent species to become dependent.

The increase in ventilation volume is effected both by a change in rate and in depth of breathing. *Salmo gairdneri* increases its rate from 80–120/min as the P_{O_2} falls from 160–40 mmHg. The increase in ventilation results in a lowered utilisation, as the increased rate of water flow across the gill lamellae reduces the time available for diffusional exchanges to proceed towards completion.

In contrast with the range of respiratory homeostasis shown by different species of fish, the circulatory response to hypoxia, though it is much less completely known, seems to be similar in the great majority of fish. Studies on both elasmobranch (Satchell, 1961, 1962; Baumgarten and Piiper, 1969) and teleost fish (Holeton and Randall, 1967*a*, *b*; Randall and Smith, 1967; Randall *et al.*, 1967) suggest that hypoxia evokes at least four circulatory adjustments. These are:

1. A reflex bradycardia accompanied by an increase in stroke volume.

2. An increase in the resistance of the branchial circulation accompanied by an increase in the transfer factor of the gills.

3. An increase in the resistance of peripheral vascular beds.

4. A partial switch to anaerobic metabolism and a rise in the level of blood lactate.

The bradycardia of hypoxia

Perfusion of *Squalus acanthias* with deoxygenated water evokes an immediate reflex bradycardia (Satchell, 1961). A similar response has been reported in *Callionymus lyra* by Hughes and Umezawa (1968). The efferent limb of the reflex is the cardiac vagus; atropine abolishes the immediate response and the heart does not slow until myocardial hypoxia supervenes after 3–4 min. The reflex occurs if the flow of blood through the gills is temporarily arrested by occluding the ventral aorta, both in *Squalus acanthias* (Satchell, 1961) and in *Salmo gairdneri* (Randall and Smith, 1967). This suggests that the chemo-receptors are accessible to the stimulus of hypoxic water and are therefore superficially located. In *Callionymus lyra* hypoxia evokes slower respirations of greater depth and a marked bradycardia. Changes in minute volume produced by altering the hydrostatic pressure across the respiratory system do not affect the heart rate (Hughes and Umezawa, 1968). This also suggests that the cardiac response is to changes in oxygen pressure rather than to flow rate. The location of possible chemoreceptors in the pseudobranch, and branchial arches was discussed in Chapter 7.

In *Scyliorhinus stellaris* Baumgarten and Piiper (1969) report that the bradycardia is accompanied by a fall in cardiac output. Decreasing the P_{O_2} of the inspired water from 140 to 60 mmHg reduced the cardiac output from 28–6 ml/kg/min. In *Salmo gairdneri* Holeton and Randall (1967a) could detect no significant reduction in cardiac output; as the heart rate decreased from 72–22/min the stroke volume correspondingly increased.

Hypoxic bradycardia occurs in fish that venture into air from time to time. The common eel can crawl considerable distances out of water and periodically fills its gill cavity with air. The heart slows as it emerges on to land (Berg and Steen, 1965). The air-breathing fish *Symbranchus marmoratus* exhibits an initial bradycardia when it leaves the water (Johansen, 1966). The grunion, *Leuresthes tenuis*, is one of few fish that lay their eggs on land. When it leaves the water to dig into the sand its heart slows from more than 100 to less than 10/min (Garey, 1962). The heart of the Californian flying fish, *Cypsilurus californicus*, slows as it emerges from the water. Two known exceptions are also of interest. The hagfish *Eptatretus* does not exhibit a hypoxic bradycardia; we have already noted that the heart is aneural, so no such reflex is to be expected. The mud skipper, *Periophthalmodon australis*, is a fish uniquely adapted to an aerial existence in swamps; when alarmed it retreats to a burrow in the mud. When forcibly submerged it exhibits a bradycardia like that of a diving amphibian (Garey, 1962). The level of oxygen pressure at which brady-cardia appears varies with the species of fish. In tench, Randall and Shelton (1965) found that the heart slowed only when the

P_{O_2} of the inspired water fell below 45 mmHg. In the rainbow trout the reflex appears at 80–100 mmHg (Randall and Smith, 1967). These authors suggest that the precise pressure at which bradycardia appears may be related to the P_{O_2} required to produce desaturation of the blood. The P_{50} of the blood of cyprinid fish is usually below that of salmonid fish.

The response of the gill vessels

In *Scyliorhinus stellaris* and *Salmo gairdneri* Baumgarten and Piiper (1969) calculated the change in vascular resistance in the gills by dividing the drop in pressure between the ventral and dorsal aortae by the cardiac output. They showed that during hypoxia vasoconstriction occurs in the gill vessels. This conclusion had previously been suggested by Satchell (1962) in *Squalus acanthias*; hypoxia caused the blood pressure in the ventral aorta to rise as that in the dorsal aorta fell; simultaneously the opacity of the gills diminished as would be expected if their lamellae contained less blood. It was also demonstrated that an element of this response remained after denervating the gills, and it was concluded that the constriction may result from the intrinsic response of contractile elements in the gill vasculature. Randall *et al.* (1967) have determined the transfer factor of the gills of *Salmo gairdneri* under normoxic and hypoxic conditions. The transfer factor expresses the ability of the respiratory surface to exchange gases and is calculated by dividing the oxygen uptake in ml/min by the mean gradient of P_{O_2} between the water and the blood, in mmHg. Lowering the P_{O_2} of the inspired water from 140 mmHg to 40 mmHg increased the transfer factor by almost four. We are not yet in a position to say what elements in the gill vasculature respond to hypoxia. An increase of resistance implies a contraction rather than a relaxation of some vascular channel. It may well be that the fibrous elements visible in electron-photomicrographs of the pillar cells of the gill lamellae(Hughes and Grimstone, 1965) are contractile and can diminish the thickness of the lamellae. This might be expected to enhance the transfer factor by diminishing the distance between the outside surface and the centre of the lamella, thereby shortening the path down which oxygen must diffuse. It would also, presumably, increase the resistance to blood flow.

93

The response of the peripheral vessels

The gradient of the descending slope of the dorsal aortic pressure pulse diminishes during hypoxia and this led Satchell (1962) in *Squalus acanthias* and Holeton and Randall (1967a) in *Salmo gairdneri* to suggest that hypoxia evokes a constriction of peripheral vascular beds. Baumgarten and Piiper (1969) showed that this was so in *Scyliorhinus stellaris* and *Salmo gairdneri* by calculating the peripheral resistance from the cardiac output and the drop in pressure between the dorsal aorta and the central veins. In *Scyliorhinus* (Baumgarten and Piiper, 1969) and in *Squalus* (Satchell, 1962), this peripheral vasoconstriction is accompanied by a fall in dorsal aortic blood pressure. In *Salmo gairdneri* (Holeton and Randall, 1967a) the dorsal aortic pressure increases. These differences probably reflect the magnitude of the concomitant changes in cardiac output.

It is well known that in diving animals bradycardia is accompanied by vasoconstriction of muscle beds and that the muscle tissue shifts to anaerobic function. It is believed that oxygen-demanding tissues such as the heart and brain are selectively supplied with oxygen while the blood flow to peripheral vascular beds is restricted (Irving, 1964). We do not, at present, know whether peripheral vasoconstriction in hypoxic fish occurs in the gut or in the muscle vessels.

Changes in blood lactate

Holeton and Randall (1967b) showed that in *Salmo gairdneri* a period of hypoxia increased the blood lactate from 12·7–34·8 mg/100 ml. A similar increase by a factor of 2–3 was reported by Garey (1962) in the grunion. In *Salmo* the increase in lactate was accompanied by a change of pH from 7·7–7·4 but this was in part due to the accumulation of carbon dioxide. At physiological pH all the lactic acid that escapes from the active muscles into the blood stream is present as lactate. Black (1955) reported that the increase of lactate caused by 15 min exercise is much greater in trout, which inhabit well-oxygenated water, than it is in carp that can survive in water of lower oxygen content. In such fish high blood lactate and the pH change it would cause, might critically depress the affinity of the haemoglobin for oxygen.

It remains to consider what are the physiological advantages of these respiratory and circulatory responses to hypoxia. Hughes (1964) and Marvin and Heath (1968) point out that the two contrasting respiratory responses of the independent and dependent species may each offer specific advantages under certain environmental conditions. Independent species like the rainbow trout, with their ability to increase ventilation to offset lowered oxygen content in the inspired water, can maintain the supply of oxygen to, and the consumption of oxygen by their tissues. In turn, this must imply that the range of behavioural responses and the speed of their execution can be sustained. Such fish are well suited to survive hypoxia of short duration. But dependent species may survive prolonged hypoxia better. Oxygen regulation may consume resources better conserved for other purposes. The reduction of oxygen consumption by dependent species may be the only course that will enable them to survive if the energy consumption involved in extracting oxygen from water with a low oxygen content is too great to be sustained.

It has been suggested that hypoxia evokes certain specific motor responses in fish. Ogden (1945) reports that if the oxygen content of inspired water is lowered, *Mustelus* become restive and exhibit swimming movements. Shepherd (1955) reports similar responses in *Salvelinus fontinalis*. We may speculate that in fish the low oxygen content of water has necessitated a haemorespiratory system in which the respiratory epithelium faces outwards into the environment. This contrasts with the mammalian condition in which the alveolar epithelium faces inwards to an enclave of alveolar air, the composition of which can be regulated by feed-back mechanisms. Perhaps in fish a reflex that links oxygen chemoreceptors and motor centres, and enables the fish to swim away from poorly oxygenated water is advantageous.

There are other observations to show that when fish are confined to hypoxic water they reduce their swimming speed as the oxygen content is lowered (Dahlberg, Shumway and Doudoroff, 1968; Kutty, 1968). It has been suggested that some reflex exists whereby enhanced chemoreceptor activity inhibits motor centres and adjusts the oxygen demands of the fish to the rate at which it can extract oxygen from the water. Both responses could be advantageous in certain situations; the problem is much in need of investigation.

The circulatory response to hypoxia can be expected to differ from those of respiration because of the difference in the oxygen affinities of blood and water. As we have seen, most fish haemoglobins are fully saturated at a partial pressure of oxygen of 40 mmHg, and have oxygen-carrying capacities of up to 14 vols%. Water in contrast, has an oxygen capacity at 10–20° C of between 0·9 and 0·6 vol% (Hughes and Shelton, 1962). At rest the ventilation rate is so adjusted that the diffusion of oxygen down the gradient that exists between the respiratory water and the blood leaving the gills is adequate to oxygenate the blood passing through them. If the oxygen content of the inspired water is reduced, the steepness of the gradiant of oxygen concentration is diminished and the rate of diffusion must fall. This can be in part compensated by increasing the ventilation rate and thereby increasing the concentration of oxygen at the more distal parts of the gill lamellae. However, viewing the haemorespiratory system as a whole, we may conclude that there is not primarily a deficiency in the transport mechanism between the gills and the tissues (Hughes and Shelton, 1962). During hypoxia of moderate severity there will be no need to increase cardiac output other than by the fractional amount necessary to supply oxygen for the additional respiratory work. The great affinity of the blood for oxygen and its great oxygen capacity compared with water implies that a fish will have to pass 15–20 vols of water over the gills for a unit volume of blood passing through them (Basu, 1959) and this ratio will increase in hypoxia. This necessarily means that the major adjustments in response to hypoxia must concern the respiratory rather than the circulatory pump (Hughes, 1964). Not until the oxygen content of the inspired water falls to a level insufficient, despite the increase in ventilation, to saturate the haemoglobin, could an increase in cardiac output enable more oxygen to be transferred to the tissues.

The increase in the transfer factor of the gills can be viewed as a necessary change that compensates in part for the diminished oxygen gradient. The peripheral vasoconstriction and accumulation of lactate presumably reflects the conservation of available oxygen by the switch to anaerobic metabolism in those tissues such as muscle and viscera that can survive for a time with this mode of respiration.

The biological advantage of the bradycardia of hypoxia has been much discussed. Heath (1964) has suggested that it serves to economise the requirements of the myocardium for oxygen. The oxygen consumption of the black grouper (*Mycteroperca bonaci*) was elevated 40–50% by vagotomy which increased the heart rate by a factor of 2·7. From this Heath calculated that the heart accounted for 26% of the resting metabolism of the fish. No account was taken of the stimulus to metabolism caused by surgery. Holeton and Randall (1967b) could find no significant reduction in cardiac output in *Salmo gairdneri* during hypoxia. The heart changes from a high-rate low-stroke volume beat to a low-rate high-stroke volume beat. They suggest that the slower heart rate enables the blood to remain longer in the gill lamellae, and that the greater time for equilibration may offset the diminished gradient of oxygen pressure between water and blood. The observation that bradycardia coincides with the onset of cardiorespiratory synchrony suggests that there may be some spatial and temporal alignment of blood and water flow that increases the transfer factor. At the present time, we must conclude that the significance of the bradycardia of hypoxia in fish remains obscure. The cardiac response to hypoxia in mammals contrasts with that in fish; hypoxia accelerates the mammalian heart and elevates the blood pressure. Daly and Scott (1958) have shown in the dog, that when the hyperpnoea of hypoxia is prevented by artificial positive pressure ventilation, perfusion of the carotid body chemoreceptors with venous blood from a donor dog causes bradycardia. Moreover, in dogs spontaneously breathing room air, denervation of the lungs converts the cardiac acceleration, which is the normal response to such perfusion, to a bradycardia, despite the hyperpnoea. They suggest that the cardioacceleratory response to hypoxia in mammals is a vagally relayed reflex from the lungs depending on hyperpnoea and masking an underlying bradycardia. The response in fish may thus be regarded as the basic vertebrate response to hypoxia which in most mammals is overridden by a phylogenetically more recent reflex related to aerial respiration.

The response to carbon dioxide

Increasing the carbon dioxide concentration in the inspired water of a tench (Randall and Shelton, 1965) causes an increase of heart rate. This is not due to cardiorespiratory synchrony, as respiration decreases in rate and increases in depth. In general the amplitude of breathing movements becomes maximal and remains high in carbon dioxide concentrations of 25–75 mg/l and there is a net increase in respiratory volume. The degree of saturation of the blood with oxygen is depressed by carbon dioxide, and an increase in cardiac output would tend to compensate for this. Saunders (1962) has suggested that the increased carbon dioxide depresses the percentage utilisation of oxygen in fish, because it depresses the affinity of haemoglobin for oxygen.

CHAPTER 10

The Circulatory Response to Exercise

So much of the data concerning the cardiovascular physiology of fishes have been derived from anaesthetised restrained specimens that it is still difficult to formulate an account of the response to exercise. However, the studies of Hanson (1967) on *Squalus*, *Raja* and *Hydrolagus*, and of Johansen, Franklin and Van Citters (1966) on *Squalus*, *Raja* and *Heterodontus* were made on unrestrained fish. Similar studies on unrestrained teleost fish have been reported for the genera *Lepomis*, *Ictalurus* and *Salmo trutta* (Sutterlin, 1969). We know most, however, about the rainbow trout, as *Salmo gairdneri* has been the species of choice both in the more recent respirometer studies of Randall and Smith (1967), Randall *et al.* (1967), Stevens and Randall (1967*a*, *b*), Stevens (1968) and on the more classical studies of Black (1955), Black *et al.* (1960, 1962) and Stevens and Black (1966).

The response of the heart

When *Salmo gairdneri* is forced to swim at a speed of 0·54 m/sec for 15 min, oxygen consumption, ventilation, and cardiac output all increase approximately fivefold. The rise in cardiac output is achieved by a small increase in rate from 47–54/min, and a large increase in stroke volume from 0·17–0·68 ml (Stevens and Randall, 1967*b*). It is accompanied by an increase in blood pressure in both the ventral and dorsal aorta. This has been observed also in *Oncorhynchus nerka* (Randall and Stevens, 1967) and *Squalus* and *Raja* (Hanson, 1967).

The response of the peripheral vessels

Exercise evokes an overall reduction of peripheral resistance. Stevens (1968) has calculated that in *Salmo gairdneri* the peripheral vessels, excluding the gill vasculature, exhibit a reduction of resistance from 142 to 41 pr units, a decrease of 71%.

99

Most of this reduction must occur within the white muscle as it constitutes 66% of the wet weight of the fish. Stevens and Black (1966) have suggested that this tissue may be equipped with vascular shunts; in *Heterodontus* Satchell (1965) reported that blood flow through the isolated perfused trunk was increased following electrically evoked swimming movements. We do not yet know whether fish possess a mechanism whereby exercise causes the liberation of dilator metabolites such as occurs in higher vertebrates.

It appears that visceral vessels are constricted during exercise. Stevens and Randall (1967a) reported that pressure in the subintestinal vein of *Salmo gairdneri* rose from $12 \cdot 1 \pm 1 \cdot 3$ to $24 \cdot 3 \pm 7 \cdot 7$ cmH$_2$O; simultaneously flow was reduced. Perhaps there is a constriction of vessels in the liver; in elasmobranch fish the sphincters that guard the entrance of the hepatic veins into the sinus venosus (Johansen and Hanson, 1967) may constrict. Such a response would tend to ensure that the augmented blood pressure enhanced the perfusion of somatic rather than visceral vessels. Stevens (1968) has reported that in *Salmo gairdneri* there is a 40% reduction in the blood volume of the spleen, suggesting a constriction of the smooth muscle of this organ. This may, in part, account for the increase in haemoglobin content of the blood during exercise reported by Black *et al.* (1962); movement of water from the blood into the tissues may also be responsible for this.

The response of the gill vessels

In *Salmo gairdneri* exercise causes a decrease in the resistance of the gill vessels of 68%, from 71 to 23 pr units (Stevens, 1968). This dilatation is accompanied by reduction in the ratio of ventral aortic to dorsal aortic pressure. These changes suggest that a relaxation of vascular smooth muscle opens additional channels to blood flow; it is accompanied by a fivefold increase in the transfer factor of the gills. Five times as much oxygen diffuses from the water to the blood per unit of oxygen concentration difference. This suggests that there is an increase in the effective area over which diffusion can occur.

Control of the area of respiratory surface perfused with blood could operate at the three anatomical levels of (*a*) the individual secondary lamella, (*b*) the gill filament, and (*c*) the entire gill.

Within each secondary lamella relaxation of branchial arterioles may distribute blood to channels more favourably placed for gas exchange, as was discussed in Chapter 4. In each particular gill, some filaments that are largely shut down at rest may be perfused during exercise. Davis (1969) finds that in quiescent *Salmo gairdneri* the P_{O_2} of the expired water varies with the placement of the cannula behind the operculum used to collect the sample. Placements which collect samples efferent to the more dorsally located filaments often have a higher P_{O_2} than water sampled more ventrally at rest. Gill filaments close to the ventral aorta may be more vigorously perfused with blood than those more dorsally placed. Overall changes in the vascular resistance of entire gills may serve to shunt blood preferentially through one rather than another. Piiper and Schumann (1967) found that the P_{O_2} of the expired water in the separate gill slits of *Scyliorhinus stellaris* differed notably. Satchell, Hanson and Johansen (1970) report that in *Raja rhina* blood flow through the common arterial stem that supplies the third, fourth and fifth gill, increases during swimming, compared with flow through the vessel supplying the first and second gill.

Baumgarten *et al.* (1969) report that in *Salmo gairdneri* there is an increase in the outward flux of sodium ions from the gills during activity doubtless related to the increased flow of blood through them. The addition of noradrenaline to the water further increased this flux. These results suggest that the functional exchange area of the gills requires to be delicately adjusted to the conflicting claims of osmoregulation and gas exchange. The gills provide a surface for the exchange of both ions and water as well as respiratory gases and it may be necessary to restrict the area of this surface at rest in order to minimise osmoregulatory work. The various species of tunny such as *Thunnus thynnus* and *Katsuwonus pelamis* have relatively enormous gill areas (Muir and Hughes, 1969) and this may explain why such fish are limited to marine environments.

Changes in the blood

In *Salmo gairdneri* Stevens and Randall (1967*b*) report that exercise causes no significant change in the P_{O_2} of either arterial or venous blood, or expired water. It appears that the increase in ventilation and cardiac output precisely matches the increased demands of the tissues for oxygen. The P_{CO_2} of the venous blood rose from 5·7 to 8 mmHg; that of the arterial blood from 2·3 to 3·3 mmHg.

Coincident with the onset of exercise there is an immediate rise in muscle lactate and a fall in muscle glycogen. Moderate exercise depletes the muscle glycogen of *Salmo gairdneri* by 50% within the first two minutes (Black *et al.* 1960, 1962). Resting muscle lactate rises from 66–295 mg/100 g after 15 min. However, this store of lactate within the muscle escapes only slowly into the blood, so that the level of blood lactate continues to rise for two hours after the cessation of a brief bout of vigorous exercise, and remains elevated for 12–24 hr. Similarly the muscle glycogen is restored slowly and has achieved only half the resting level after 24 hr.

The release of muscle lactate into the blood increases the hydrogen ion concentration; Black (1958) reports that in carp, exercise changes the pH from 7·33–7·25, and the bicarbonate falls from 35 vol% to 17·5 vol%. Some of the blood lactate is converted in the liver to glucose and glycogen (Black, 1955). Some of it is excreted in the urine. Liver glycogen is not drawn upon during exercise; it is however depleted by starvation.

In Chapter 4 it was pointed out that whereas white muscle constituted 66% of the body weight and red muscle 1%, yet it had only one-third of the capillary volume of red muscle. Its poor vascularisation is in accord with the finding that it becomes active only during vigorous swimming and functions glycolytically converting glycogen to lactic acid which is then liberated slowly into the blood stream. The brief periods of vigorous swimming which are all that the majority of teleost fish can sustain must be followed by prolonged periods of rest during which either the liver reconverts the lactic acid back into glycogen or the lactic acid is excreted. Fish living in poorly oxygenated water face the problem that the rise of plasma hydrogen ion concentration caused by the liberation of muscle lactate may critically depress the affinity of the haemoglobin for oxygen and interfere with its uptake at the gills. Fish such as carp and catfish have bloods with a small Bohr effect; they are also languid swimmers which do not exhibit intense bursts of activity. These characterise fish like trout and salmon, which inhabit well-oxygenated waters.

Presumably the rate at which the white muscle liberates its lactate into the blood stream following exercise must be finely adjusted to the sensitivity of the haemoglobin to pH changes in different species of fish.

The role of intrinsic and extrinsic responses in the adjustments to exercise

The existence of venous pumps operated by movements of trunk, fins and tail was outlined in Chapter 3. Satchell (1965) showed that electrically evoked swimming movements increased venous outflow from the perfused post-pelvic trunk of *Heterodontus* by 46%. Sutterlin (1969) reported that flexing the tail of *Lepomis* and *Ictalurus* increased the pressure in the caudal vein, distended the central veins, and increased the heart rate. These mechanisms coupled with the dilation of muscle vessels will serve to increase both venous return and venous filling pressure and this, we may suppose, will serve to augment both stroke volume and heart rate by the two intrinsic responses of the myocardium outlined in Chapter 8.

Nakano and Tomlinson (1967) report that exercise brings about an increase in the level of circulating catecholamines in *Salmo gairdneri*. These agents are known to be present both in the innervated chromaffin tissue, and in adrenergic nerve terminals. Their liberation into the blood stream can be expected to evoke responses in all organ systems that possess appropriate receptors. Falck *et al.* (1966) have demonstrated the existence of beta receptors in the isolated heart of the plaice (*Pleuronectes platessa*) and adrenaline has repeatedly been observed to have both ino- and chronotropic effects on the fish heart, and to augment its output. Östlund and Fänge (1962) have shown that both adrenaline and noradrenaline diminish the vascular resistance of isolated fish gills. Kirby and Burnstock (1969) report that adrenaline evokes contraction of isolated strips of the ventral aorta of the trout and eel. It is thus likely that catecholamines, whether liberated from central or peripheral storage sites, effect an increase of cardiac output, an increase in the transfer factor of the gills by dilation of specific vascular channels, and a redistribution of blood flow to muscle by vasoconstriction of specific vascular beds such as the gut.

We are at present largely ignorant of the relative importance of the control exercised by the autonomic nervous system through the central liberation of circulating catecholamines and the direct action of post-ganglionic fibres to specific vascular structures. Stevens and Randall (1967a) report that in *Salmo gairdneri* doses of atropine sufficient to block the cholinergic vagal fibres to the heart neither change the resting heart rate nor inhibit the increase in heart rate during exercise. They concluded that in this species, the release of vagal tone plays no part in the increase in cardiac output. In the sucker (*Catostomus macrocheilus*) atropine increases the resting heart rate by as much as 75% but does not increase the maximum rate observed during exercise. They concluded that in this species there is a resting vagal tone, the reduction of which is in part responsible for the increase in heart rate.

Rodionov (1959) has shown the existence of vasomotor fibres to the gut in pike and other teleost fish. Mechanical stimulation of the gills reflexly evoked changes in the outflow of blood from intestinal veins. Randall and Stevens (1967) showed that in the salmon, *Oncorhynchus nerka* and *O. kisutch*, exercise causes the blood pressure to increase by 13·5 and 20·2 cmH$_2$O. The alpha blocking drug, phenoxybenzamine, changes this rise to a slight fall in pressure. This is consistent with the blocking of vasoconstrictor responses in visceral vessels, but it does not tell us whether these were mediated by post-ganglionic fibres or circulating catecholamines. The problem is urgently in need of further investigation.

The question of what are the stimuli which initiate the respiratory and circulatory adjustments to exercise in fish has been discussed by various workers. The question needs to be preceded by a more fundamental one; what parameter of the haemorespiratory system is it expedient for the fish to monitor if it is to detect and respond to changes in the oxygen consumption. The question has been discussed by Hughes and Shelton (1962). It would seem unnecessary to monitor the P$_{CO_2}$ of the blood. It has been previously argued (Chapter 5) that the high solubility of carbon dioxide in water, compared with oxygen, makes it inherently unlikely that inadequate ventilation will primarily be signalled by a rise in blood P$_{CO_2}$. The P$_{O_2}$ of the blood on the other hand ought to be a more sensitive indication of the state of balance between oxygen supply and oxygen demand: it changes considerably if access to oxygen is hindered, even for a short period. Computer-simulation studies of cardiovascular regulation in salmonid fish (Taylor, Houston and Horgan, 1968) strongly support the view that respiratory responses are based upon monitoring blood P_{O_2}. Hughes (1964) suggested that the monitor was on the arterial side; Taylor *et al.* (1968) found better agreement with their computer-simulation, and the data of Stevens and Randall (1967*a*, *b*) and Randall *et al.* (1967) if the monitor was located on the venous side. The problem remains obscure, however, because in *Salmo gairdneri* the observed changes in the P_{CO_2} and P_{O_2} of the arterial and venous blood are not of sufficient magnitude to account for the increase in ventilation and cardiac output (Stevens and Randall, 1967*b*). The adjustments of these appear to be of such a magnitude that they maintain the blood gas concentrations approximately constant despite the fivefold increase in resting oxygen consumption.

In mammals it is widely believed that receptors in joints and tendons, activated by movement, stimulate the respiratory centre. It would seem that in fish too it is necessary to postulate the existence of some additional source of respiratory drive derived from receptors stimulated by swimming movements. In *Scyliorhinus* Roberts (1969) has shown that the significant sensory input utilised in the production of co-ordinated loco-motory movements is derived from the skin. In fish the skin is very closely attached to the muscles and moves when they move. If it transpires that in fish the hyperventilation of exercise is indeed driven from the input of some peripheral receptors activated by swimming, then the possibility that these are cutaneous receptors will need to be considered. It has long been known that the heart and respiration of fish are very susceptible to inhibition by quite trivial cutaneous stimulation (Kisch, 1950).

The circulatory response to exercise can, it has been argued, be in part ascribed to the increase in venous return caused by movement, and the dilation of muscle vessels caused by the local accumulation of metabolites. In addition we know that there are reflex influences brought to bear on the heart from the respiratory system; cardiorespiratory synchrony is an example of this. In mammals the circulatory autonomic centres are powerfully stimulated by collateral fibres coming from the motor areas of the brain. Thus motor impulses not only effect movements but also stimulate the diencephalic and medullary vasomotor centres. The possibility that an analogous linkage of motor and circulatory centres occurs in the fish brain needs to be investigated.

CHAPTER 11

Future Developments

The tempo of research into the circulatory physiology of fish has quickened in the last decade; 40% of the papers cited in this book have been published in or after 1965. This acceleration will undoubtedly continue because we are now entering an era in which more sophisticated recording techniques are becoming available to the comparative physiologist. Our ability to measure the partial pressure of the respiratory gases by electrometric methods, and the increasing miniaturisation of electrodes will enable gas transport to be investigated more precisely. The development of the technique of stimulating and recording from the nervous system through implanted electrodes offers hope for a better understanding of the regulation of the circulatory system and the part that circulatory reflexes play in this. The use of telemetric recording whereby ultrasonic sound is modulated by the output of implanted transducers is an exciting field of great promise. It will permit the study of larger fish than those which can at present be persuaded to swim within the confines of a respirometer.

Circulation in fish is only one aspect of the wider field of circulatory physiology, and there has always been a stimulating feed-back from the one to the other. Comparative physiologists have gained greatly from ideas and insights derived from mammalian physiology (Hughes, 1966b). The regulation of the oxygen-binding power of haemoglobin by the control of the level of intracellular organic phosphates is an example of this. The mechanism was first described from clinical studies of anaemic patients and of the deterioration of stored blood in blood banks. This regulation will probably prove to be of great significance in the acclimation of fish to hypoxia and raised temperature. Hughes (1964) has emphasised that there is now a feed-back in the opposite direction, from the circulatory and respiratory physiology of fish to the wider field of mammalian studies. In fish the exchange of the respiratory gases between the environment and the blood is simpler than that in mammals because there is no buffering layer of alveolar air. The partial pressure of the respiratory gases in contact with the respiratory epithelium can thus be determined with greater precision. The quantitative aspects of gas exchange across respiratory epithelia may well be better studied in fish.

Certain aspects of the circulatory physiology of fish have an immediate practical importance. The mountains and streams of Canada and the United States constitute an environment suitable for the commercial exploitation of salmon and timber. Unfortunately, the effluents of paper mills lower the oxygen content of stretches of rivers that may be crucial highways in the spawning migrations of these fish. Data on the acclimation of fish to hypoxia may be highly relevant to those concerned with formulating the management policies of such river systems.

The expanding world population, and the widespread lack of protein in the diet of many developing countries suggest that there must be an increasing harvest of fish from natural and artificial waters. A recent study by F.A.O. (1968) suggests that the world catch of fish of 52 million tons in 1965 will have to increase by 24–38 million tons by 1985 if the additional requirements due to population growth are to be met. If protein undernutrition in the world's population were to be eliminated by the year 2000, the present world production of animal protein would need to increase by 43·8%. In the Far East region alone, which includes mainland China, this would require a growth of the fish catch from the 1960 level of 14 million tons to 82 million tons in the year 2000. Whilst some of this increase may be achieved by the discovery of new resources, much more must depend on the effective management of existing stocks and the artificial cultivation of fish. We know a great deal of the physiology of our domestic meat-producing mammals and such knowledge has contributed to the high efficiency of commercial meat production. It seems unlikely that a comparable efficiency in the production of fish will be achieved without a comparably detailed understanding of their physiology.

REFERENCES

Akerblom, O., De Verdier, C. H., Garby, L. and Högman, C. (1968). Restoration of defective oxygen transport function of stored red blood cells by addition of inosine. *Scand. J. clin. Lab. Invest.* **21,** 245–8.

Albers, J. A. A. (1806). Über das Auge des Kabeljau *Gadus morhua,* und die Schwimmblase der Secschwalb, *Trigla hirundo. Götting gelehrte Anz.* **2,** 681–2.

Alexander, R. M. (1966). Physical aspects of swimbladder function. *Biol. Rev.* **41,** 141–76.

Allen, W. F. (1949). Blood vascular system of the eye of a deep water fish (*Ophiodon elongatus*) considered as a pressure mechanism. *Anat. Rec.* **103,** 205–11.

Anthony, E. H. (1961). The oxygen capacity of goldfish blood in relation to thermal environment. *J. Exp. Biol.* **38,** 93–107.

Augustinsson, K. B., Fänge, R., Johnels, A. and Östlund, E. (1956). Histological, physiological, and biochemical studies on the heart of two cyclostomes, Hagfish (*Myxine*) and Lamprey (*Lampetra*). *J. Physiol., Lond.* **131,** 257–76.

Ball, E. G., Strittmatter, C. F. and Cooper, O. (1955). Metabolic studies on the gas gland of the swim bladder. *Biol. Bull. mar. biol. Lab., Woods Hole,* **108,** 1–17.

Barnett, C. H. (1951). The structure and function of the choroidal gland of teleostean fish. *J. Anat.* **85,** 113–19.

Barrett, I. and Connor, A. R. (1962). Blood lactate in yellow fin tuna *Neothunnus macropterus* and skipjack, *Katsuwonus pelamis* following capture and tagging. *Int. trop. Tuna Comm.* **6,** 233–80.

Barrett, I. and Tsuyuki, H. (1967). Serum transferrin polymorphism in some Scombroid fishes. *Copeia.* **3,** 551–7.

Basu, S. P. (1959). Active respiration of fish in relation to ambient concentrations of oxygen and carbon dioxide. *J. Fish. Res. Bd Can.* **16,** 175–212.

Baumgarten, D. and Piiper, J. (1969). Effects of hypoxia upon respiration and circulation in the dogfish *Scyliorhinus stellaris.* (In the press.)

Baumgarten, D., Randall, D. J. and Malyusz, J. (1969). Gas exchange versus ion exchange across the gills of fish. (In the press.)

Beaumont, E. and Randall, J. D. (1969). Unpublished data.

Becker, E. L., Bird, R., Kelly, J. W., Schilling, J., Solomon, S. and Young, N. (1958). Physiology of marine teleosts. 2. Hematologic observations. *Physiol. Zoöl.* **31,** 228–31.

Benesch, R. and Benesch, R. E. (1967). The effect of organic phosphates from the human erythrocyte on the allosteric properties of haemoglobin. *Biochem. Biophys. Res. Commun.* **26,** 162.

Benesch, R. and Benesch, R. E. (1969). Intracellular organic phosphates as regulators of oxygen release by haemoglobin. *Nature, Lond.* **221,** 618–22.

Bennett, H. S., Luft, J. H. and Hampton, J. C. (1959). Morphological classifications of vertebrate blood capillaries. *Am. J. Physiol.* **196,** 381–90.

Benninghoff, A. (1933). Das Herz. In *Vergleichende Anatomie*. Ed. by Bolk, L., Goppert, E., Kallius, E. and Lubosch, W. **6**, 467–556. Berlin: Urban and Schwarzenberg.

Bennion, G. R. and Randall, D. J. (1970). Starling's law and the control of the isolated, perfused trout heart. (In the press.)

Berg, T. and Steen, J. B. (1965). Physiological mechanisms for aerial respiration in the eel. *Comp. Biochem. Physiol.* **15**, 469–84.

Berg, T. and Steen, J. B. (1968). The mechanism of oxygen concentration in the swim bladder of the eel. *J. Physiol., Lond.* **195**, 631–8.

Birch, M. P., Carre, C. G. and Satchell, G. H. (1969). Venous return in the trunk of the Port Jackson Shark, *Heterodontus portusjacksoni*. *J. Zool.* **159**, 31–49.

Black, E. C. (1940). The transport of O_2 by the blood of fresh water fish. *Biol. Bull. mar. biol. Lab.*, *Woods Hole*, **79**, 215–29.

Black, E. C. (1955). Blood levels of haemoglobin and lactic acid in some freshwater fishes following exercise. *J. Fish. Res. Bd Can.* **12**, 917–29.

Black, E. C. (1958). Energy stores and metabolism in relation to muscular activity. In *The Investigation of Fish-power Problems*. Ed. by P. A. Larkin, 51–67. Vancouver: University of British Columbia Press.

Black, E. C., Connor, A. R., Lam, K. C. and Chiu, W. G. (1962). Changes in glycogen, pyruvate and lactate in rainbow trout (*Salmo gairdneri*) during and following muscular activity. *J. Fish. Res. Bd Can.* **19**, 409–36.

Black, E. C., Kirkpatrick, D. and Tucker, H. H. (1966). Oxygen dissociation curves of the blood of brook trout (*Salvelinus fontinalis*) acclimated to summer and winter temperatures. *J. Fish. Res. Bd Can.* **23**, 1–13.

Black, E. C., Robertson, A. C., Hanslip, A. R. and Chiu, W. G. (1960). Alteration in glycogen, glucose and lactate in rainbow and Kamloops trout *Salmo gairdneri* following muscular activity. *J. Fish. Res. Bd Can.* **17**, 487–500.

Black, E. C., Tucker, H. H. and Kirkpatrick, D. (1966). Oxygen dissociation curves of the blood of atlantic salmon, *Salmo salar*, acclimated to summer and winter temperatures. *J. Fish. Res. Bd Can.* **23**, 1187–95.

Blinks, J. R. (1956). Positive chronotropic effect of increasing right atrial pressure in the isolated mammalian heart. *Am. J. Physiol.* **186**, 299–303.

Booke, H. E. (1964). A review of variations found in fish serum proteins. *J. New York Fish. and Game*, **11**, 47–51.

Boyd, J. D. (1936). Nerve supply to the branchial arch arteries of vertebrates. *J. Anat.* **71**, 157–8.

Brady, A. J. and Woodbury, J. W. (1960). The sodium-potassium hypothesis as the basis of electrical activity in frog ventricle. *J. Physiol., Lond.* **154**, 385–407.

Braunitzer, G. and Hilse, K. (1963). Zur Phylogenie des Hämoglobin-Moleküls. Die Konstitution des Karpfenhamoglobins. *Hoppe-Seyler's Z. physiol. Chem.* **330**, 234–6.

Brunings, W. (1899). Zur Physiologie des Kreislaufes der Fische. *Pflugers Arch. ges. physiol.* **75**, 599–641.

Bugge, J. (1960). The heart of the African lungfish, *Protopterus*. *Vidensk, Meddr dansk naturh. Foren.* **123**, 193–210.

Buhler, D. R. and Shanks, W. E. (1959). Multiple hemoglobins in fishes. *Science, N.Y.* **129**, 899–900.

Bulger, R. E. and Trump, B. J. (1969). Ultrastructure of granulated arteriolar cells (juxtaglomerular cells) in kidney of a fresh and a salt water teleost. *Am. J. Anat.* **124**, 77–88.

Burger, J. W. and Bradley, S. E. (1951). The general form of the circulation in the dogfish *Squalus acanthias*, *J. Cell. comp. physiol.* **37**, 389–402.

Burne, R. H. (1909). On elastic mechanisms in fishes and a snake. *Proc. zool. Soc. Lond.* 201–3.

Burnstock, G. (1969). The evolution of autonomic nervous control of the visceral and cardiovascular system. (In the press.)

Carey, F. G. and Teal, J. M. (1969a). Mako and Porbeagle; warm blooded sharks. *Comp. Biochem. Physiol.* **28**, 199–204.

Carey, F. G. and Teal, J. M. (1969b). Regulation of body temperature by the blue fin tuna. *Comp. Biochem. Physiol.* **28**, 205–13.

Casley-Smith, J. R. and Hart, P. R. (1970). The relative antiquity of fenestrated blood capillaries and lymphatics and their significance for the uptake of large molecules; an electron microscopical investigation in an elasmobranch. (In the press.)

Chandrasekhar, N. (1959). Multiple haemoglobins in fish. *Nature, Lond.* **184**, 1652–3.

Chapman, C. B., Jensen, D. and Wildenthal, K. (1963). On circulatory control mechanisms in the Pacific hagfish. *Circulation Res.* **12**, 427–40.

Chiesa, D. F., Noseda, V. and Marchetti, R. (1962). Attivazione degli strati epicardici nel cuore de alcuni teleosti di acqua dolce. Ricerche elettrocardiografiche. *Archo. Sci. Biol.* **46**, 1–10.

Conte, F. P., Wagner, H. H. and Harris, T. O. (1963). Measurement of blood volume in the fish *Salmo gairdneri*. *Am. J. Physiol.* **205**, 533–40.

Conteaux, R. and Laurent, P. (1958). Observations au microscope electronique sur l'innervation cardiaque des téléostéens. *Bull. Ass. Anat., Paris,* **98**, 230.

Copeland, D. E. and Dalton, A. J. (1959). An association between mitochondria and the endoplasmic reticulum in cells of the pseudobranch gland of a teleost. *J. biophys. biochem. Cytol.* **5**, 393–5.

Dahlberg, M. L., Shumway, D. L. and Doudoroff, P. (1968). Influence of dissolved oxygen and carbon dioxide on swimming performance of large mouth bass and coho salmon. *J. Fish. Res. Bd Can.* **25**, 49–70.

Daly, M. De Burgh and Scott, M. J. (1958). The effects of stimulation of the carotid body chemoreceptors on heart rate in the dog. *J. Physiol., Lond.* **144**, 48–166.

Davis, J. (1969). Personal communication.

De Kock, L. L. (1963). A histological study of the head region of two salmonids with special reference to pressure and chemoreceptors. *Acta anat.* **55**, 39–50.

De Kock, L. L. and Symmons, S. (1959). A ligament in the dorsal aorta of certain fishes. *Nature, Lond.* **184**, 194.

De Marco, C. and Antonini, E. (1958). Amino acid composition of haemoglobin from *Thunnus thynnus*. *Nature, Lond.* **181**, 1128.

Denton, E. J. (1961). The buoyancy of fish and cephalopods. *Prog. Biophys. biophys. Chem.* **11**, 178–234.

Deutsche, H. F. and McShan, W. H. (1949). Biophysical studies of blood plasma proteins. XII. Electrophoretic studies of the blood serum proteins of some lower animals. *J. biol. Chem.* **180**, 219–34.

De Wilde, M. A. and Houston, A. H. (1967). Haematological aspects of the thermoacclimatory process in the rainbow trout, *Salmo gairdneri*. *J. Fish. Res. Bd Can.* **24**, 2267–81.

Dill, D. B., Edwards, H. T. and Florkin, M. (1932). Properties of the blood of the skate (*Raia oscillata*). *Biol. Bull. mar. biol. Lab., Woods Hole*, **62**, 23–56.

Doolittle, R. F. (1963). Further studies on clotting and fibrinolysis in plasma from the smooth dogfish (*Mustelus canis*). *Br. J. Haemat.* **9**, 464–70.

Doolittle, R. F. and Surgenor, D. M. (1962). Blood coagulation in fish. *Am. J. Physiol.* **203**, 964–70.

Dornesco, G. T. and Santa, V. (1963). La structure des aortes et des vaisseaux de la carpe (*Cyprinus carpio* L.). *Anat. Anz.* **113**, 136–45.

Drake, E. N., Gill, S. J., Downing, M. and Malone, C. P. (1963). The environmental dependency of the reaction of oxygen with haemoglobin. *Archs. Biochem. Biophys.* **100**, 26–31.

Eguchi, H., Hashimoto, K. and Matsuura, F. (1960). Comparative studies of two haemoglobins of Salmon. III. Amino acid composition. *Bull. Jap. Soc. scient. Fish.* **26**, 810–13.

Falck, B., Mecklenburg, C. von, Myhrberg, H. and Persson, H. (1966). Studies on adrenergic and cholinergic receptors in the isolated hearts of *Lampetra fluviatilis* (Cyclostomata) and *Pleuronectes platessa* (Teleostei). *Acta physiol. scand.* **68**, 64–71.

Falkner, N. W. and Houston, A. H. (1966). Some haematological responses to sublethal thermal shock in the goldfish, *Carassius auratus* L. *J. Fish Res. Bd Can.* **23**, 1109–20.

Fänge, R. (1953). The mechanism of gas transport in the euphysoclist swim bladder. *Acta physiol. scand.* **30**. Suppl. **110**, 1–133.

Fänge, R. (1968). The formation of eosinophilic granulocytes in the eosophageal lymphomyeloid tissue in the Elasmobranchs. *Acta. zool., Stockh.* **49**, 155–61.

Fänge, R. and Wittenberg, J. B. (1958). The swimbladder of the toadfish (*Opsanus tau* L.). *Biol. Bull. mar. biol. Lab., Woods Hole*, **115**, 172–9.

Favaro, G. (1906). Ricerche intorno alla morfologia ed allo sviluppo dei vasi, seni e cuori caudali nei ciclostomi e nei pesci. *Atti. Ist. veneto Sci.* **65**, 1–279.

Field, J. B., Elvehjem, C. A. and Chancey, J. (1943). A study of the blood constituents of carp and trout. *J. biol. Chem.* **148**, 261–9.

Fish, G. R. (1956). Some aspects of the respiration of six species of fish from Uganda. *J. exp. Biol.* **33**, 186–95.

Flemming, H. (1958). Untersuchungen über die Bluteiweisskorper gesunder und bauchwassersuchtskranker Karpfen. *Z. Fisch.* **7**, 91–160.

Food and Agriculture Organisation of the United Nations (1968). Fisheries in the food economy. *Freedom from Hunger Campaign, Basic Studies*, **19** Rome.

Forster, R. E. and Steen, J. B. (1968). The rate of the Root shift of eel red cells and haemoglobin solution. *J. Physiol., Lond.* **204**, 259–83.

Gannon, J. and Burnstock, G. (1969). Excitatory adrenergic innervation of the fish heart. *Comp. Biochem. Physiol.* **29**, 765–73.

Garey, W. F. (1962). Cardiac responses of fishes in asphyxic environments. *Biol. Bull. mar. biol. Lab., Woods Hole*, **122**, 362–8.

Goldstein, L., Forster, R. P. and Fanelli, G. M. (1964). Gill blood flow and ammonia excretion in the marine teleost *Myoxocephalus scorpius*. *Comp. Biochem. Physiol.* **12**, 489–99.

Grant, R. and Regnier, M. (1926). The comparative anatomy of the cardiac coronary vessels. *Heart*, **13**, 285–317.

Greene, C. W. (1899). Contributions to the physiology of the Californian hagfish, *Polistotrema stouti* – I. The anatomy and physiology of the caudal heart. *Am. J. Physiol.* **3**, 366–82.

Greene, C. W. (1902). Contributions to the physiology of the Californian hagfish *Polistotrema stouti* – II. The absence of regulative nerves for the systemic heart. *Am. J. Physiol.* **6**, 318–24.

Greene, C. W. (1904). Physiological studies of Chinook salmon. I. Relation of blood pressure to functional activity. *Bull. U.S. Bur. Fisheries*, **24**, 431–55.

Grigg, G. C. (1967). Some respiratory properties of the blood of four species of antarctic fishes. *Comp. Biochem. Physiol.* **23**, 139–48.

Grigg, G. C. (1969). Temperature induced changes in the oxygen equilibrium curve of the blood of the brown bullhead, *Ictalarus nebulosus*. *Comp. Biochem. Physiol.* **28**, 1203–23.

Grodzinski, Z. (1954). Contractions of the isolated heart of the European Glass Eel, *Anguilla anguilla*. L. *Bull. Acad. pol. Sci. Cl. II Ser. Sci. biol.* **2**, 19–22.

Grodzinski, Z. (1964). The laminar stratification of circulating blood of the sea trout, *Salmo trutta* L., yolk sac. *Acta biol. cracov.* **7**, 183–6.

Guest, M. M., Bond, T. P., Cooper, R. G. and Derrick, J. R. (1963). Red blood cells: change in shape in capillaries. *Science, N.Y.* **142**, 1319–21.

Hanson, D. (1967). Cardiovascular dynamics and aspects of gas exchange in Chondrichthyes, Ph.D. Dissertation. University of Washington. Seattle, Washington.

Harden-Jones, F. R. and Marshall, N. B. (1953). The structure and functions of the teleostean swimbladder. *Biol. Rev.* **28**, 16–83.

Hartman, F. A. and Lessler, M. A. (1964). Erythrocyte measurements in fishes, amphibia and reptiles. *Biol. Bull. mar. biol. Lab., Woods Hole*, **126**, 83–8.

Harvey, W. (1649). De circulatione sanguinis. Another exercitation to John Riolan. Pp. 145–93. In *The anatomical exercises of Dr William Harvey*, ed. G. Keynes. London: 1949. Nonesuch Press.

Hashimoto, K., Yamaguchi, Y. and Matsuura, F. (1960). Comparative studies on two hemoglobins of salmon. IV. Oxygen dissociation curve. *Bull. Jap. Soc. scient. Fish.* **26**, 827–34.

Haws, T. G. and Goodnight, C. J. (1962). Some aspects of the hematology of two species of catfish, in relation to their habitats. *Physiol. Zoöl*, **35**, 8–17.

Heath, A. G. (1964). Heart rate, ventilation, and oxygen uptake of a marine teleost in various oxygen tensions. *Am. Zoologist*, **4**, 386.

Hemmingsen, E. A., Douglas, E. L. and Grigg, G. C. (1969). Oxygen consumption in an antarctic hemoglobin free fish, *Pagetopsis macropterus* and in three species of *Notothenia*. *Comp. Biochem. Physiol.* **29**, 467–70.

Hill, A. V. (1910). The possible effects of the aggregation of the molecules of hemoglobin on its dissociation curves. *J. Physiol., Lond.* **40**, 4–7.

Hirsch, E. F., Jellinek, M. and Cooper, T. (1964). Innervation of the systemic heart of the Californian hagfish. *Circulation Res.* **14**, 212–17.

Hoffert, J. R. (1966). Observations on ocular fluid dynamics and carbonic anhydrase in tissues of lake trout (*Salvelinus namaycush*). *Comp. Biochem. Physiol.* **17**, 107–14.

Hoffman, B. F. and Cranefield, P. F. (1960). *Electrophysiology of the Heart*. New York: McGraw-Hill.

Holeton, G. F. and Randall, D. J. (1967a). Changes in blood pressure in the rainbow trout during hypoxia. *J. exp. Biol.* **46**, 297–305.

Holeton, G. F. and Randall, D. J. (1967*b*). The effect of hypoxia upon the partial pressure of gases in the blood and water afferent and efferent to the gills of rainbow trout. *J. exp. Biol.* **46**, 317–27.

Holmgren, A. (1966). The 'oxygen conduction line' of the human body. *Proc. int. Symp. cardiovasc. respir. Effects Hypoxia*, pp. 391–400. Ed. D. Hatcher and D. B. Jennings. New York: Hafner.

Holst, R. (1969). (In preparation.)

Houston, A. H. and De Wilde, M. A. (1968*a*). Hematological correlations in the rainbow trout, *Salmo gairdneri*. *J. Fish Res. Bd Can.* **25**, 173–6.

Houston, A. H. and De Wilde, M. A. (1968*b*). Thermoacclimatory variations in the haemotology of the common carp *Cyprinus carpio*. *J. exp. Biol.* **49**, 71–81.

Houston, A. H. and De Wilde, M. A. (1969). Environmental temperature and the body fluid system of the fresh water teleost. III. Hematology and blood volume of thermally acclimated brook trout *Salvelinus fontinalis*. *Comp. Biochem. Physiol.* **28**, 877–85.

Hughes, G. M. (1960). The mechanism of gill ventilation in the dogfish and skate. *J. exp. Biol.* **37**, 11–27.

Hughes, G. M. (1964). Fish respiratory homeostasis. *Symp. Soc. exp. Biol.* **18**, 81–107.

Hughes, G. M. (1966*a*). The dimensions of fish gills in relation to their function. *J. exp. Biol.* **45**, 177–95.

Hughes, G. M. (1966*b*). Species variation in gas exchange. *Proc. roy. Soc. Med.* **59**, 494–500.

Hughes, G. M. (1966*c*). Evolution between air and water. In *Ciba-Foundation Symposium on Development of the Lung*, pp. 64–80. Ed. A. V. S. de Reuck and R. Porter. London: Churchill.

Hughes, G. M. and Ballintijn, C. M. (1965). The muscular basis of the respiratory pumps in the dogfish (*Scyliorhinus canicula*). *J. exp. Biol.* **43**, 363–83.

Hughes, G. M. and Ballintijn, C. M. (1968). Electromyography of the respiratory muscles and gill water flow in the dragonet. *J. exp. Biol.* **49**, 583–602.

Hughes, G. M. and Grimstone, A. V. (1965). The fine structure of the secondary lamellae of the gills of *Gadus pollachius*. *Q. Jl. Micr. Sci.* **106**, 343–53.

Hughes, G. M. and Roberts, J. L. (1970). A study of the effect of temperature changes on the respiratory pumps of the rainbow trout. *J. exp. Biol.* **52**, 177–92.

Hughes, G. M. and Shelton, G. (1958). The mechanism of gill ventilation in three freshwater teleosts. *J. exp. Biol.* **35**, 807–23.

Hughes, G. M. and Shelton, G. (1962). Respiratory mechanisms and their nervous control in fish. *Advance Comp. Physiol. Biochem.* **1**, 275–364.

Hughes, G. M. and Umezawa, S. I. (1968). On respiration in the dragonet, *Callionymus lyra* L. *J. exp. Biol.* **49**, 565–82.

Hughes, G. M. and Wright, D. E. (1970). A comparative study of the ultrastructure of the water–blood pathway in the secondary lamellae of teleost and elasmobranch fishes—Benthic forms. *Z. Zellforsch.* **104**, 478–93.

Irisawa, H. and Irisawa, A. F. (1954). Blood serum protein of the marine Elasmobranchii. *Science, N.Y.,* **120**, 849–50.

Irving, L. (1964). Comparative anatomy and physiology of gas transport mechanisms. In *Handbook of Physiology*, Section 3. Respiration, **1**, 177–212. Ed. W. O. Fenn and H. Rahn. Washington, D.C.: American Physiological Society.

Irving, L., Solandt, D. V. and Solandt, O. M. (1935). Nerve impulses from branchial pressure receptors in the dogfish. *J. Physiol., Lond.* **84**, 187–90.

Jakubowski, M. (1960). The structure and vascularization of the skin of the eel and the viviparous blenny. *Acta biol. cracov.* **3**, 1–22.

Jakubowski, M. (1963). The structure and vascularization of the skin of the river bull head (*Cottus gobio* L.) and Black Sea turbot (*Rhombus maeoticus* Pall). *Acta biol. cracov.* **6**, 159–75.

Jensen, D. (1961). Cardio regulation in aneural hearts. *Comp. Biochem. Physiol.* **2**, 181–201.

Jensen, D. (1963). Eptatretin; a potent cardioactive agent from the branchial heart of the pacific hagfish *Eptatretus stoutii. Comp. Biochem. Physiol.* **10**, 129–51.

Jensen, D. (1965). The aneural heart of the hagfish. *Ann. N.Y. Acad. Sci.* **127**, 443–58.

Jesse, M. J., Shub, C. and Fishman, A. P. (1967). Lung and gill ventilation in the African lung fish. *Resp. Physiol.* **3**, 267–87.

Johansen, K. (1960). Circulation in the hagfish, *Myxine glutinosa* L. *Biol. Bull. mar. biol. Lab.*, *Woods Hole*, **118**, 289–95.

Johansen, K. (1962). Cardiac output and pulsatile aortic flow in the teleost *Gadus morhua. Comp. Biochem. Physiol.* **7**, 169–74.

Johansen, K. (1963). The cardiovascular system of *Myxine glutinosa.* In *The Biology of Myxine*, pp. 289–316. Ed. A. Brodal and R. Fänge. Oslo: Universitets Forlaget.

Johansen, K. (1965a). Cardiovascular dynamics in fishes amphibians and reptiles. *Ann. N.Y. Acad. Sci.* **127**, 414–42.

Johansen, K. (1965b). Dynamics of venous return in elasmobranch fishes. *Hvalrad Skr.* **48**, 94–100.

Johansen, K. (1966). Air breathing in the teleost, *Symbranchus marmoratus. Comp. Biochem. Physiol.* **18**, 383–95.

Johansen, K., Franklin, P. L. and Van Citters, R. L. (1966). Aortic blood flow in free swimming elasmobranchs. *Comp. Biochem. Physiol.* **19**, 151–60.

Johansen, K. and Hanson, D. (1967). Hepatic vein sphincters in elasmobranchs and their significance in controlling hepatic blood flow. *J. exp. Biol.* **46**, 195–203.

Johansen, K. and Hanson, D. (1968). Functional anatomy of the hearts of lungfishes and amphibians. *Am. Zoologist*, **8**, 191–210.

Johansen, K. and Lenfant, C. (1967). Respiratory function in the south American lungfish, *Lepidosiren paradoxa* (Fitz.). *J. exp. Biol.* **46**, 205–218.

Johansen, K. and Lenfant, C. (1969). Respiration in the African lungfish *Protopterus aethiopicus*: control of breathing. *J. exp. Biol.* **49**, 453–68.

Johansen, K., Lenfant, C. and Grigg, G. C. (1967). Respiratory control in the lungfish *Neoceratodus forsteri* (Krefft). *Comp. Biochem. Physiol.* **20**, 835–54.

Johansen, R., Lenfant, C. and Hanson, D. (1968a). Cardiovascular dynamics in the lungfishes. *Z. vergl. Physiol.* **59**, 157–86.

Johansen, K., Lenfant, C. and Schmidt-Nielsen, K. (1968b). Gas exchange and control of breathing in the electric eel *Electrophorus electricus. Z. vergl. Physiol.* **61**, 137–63.

Johansen, K. and Martin, A. W. (1965). Comparative aspects of cardiovascular function in vertebrates. In *Handbook of Physiology*, Section 2. Circulation, **3**, 2583–614. Ed. W. F. Hamilton. Washington, D.C.: American Physiological Society.

Jullien, A. and Ripplinger, J. (1957). Physiologie du coeur des poissons et son innervation extrinsèque. *Annls. scient. Univ. Besancon Zool. et Physiol.* **9**, 35–92.

Keatinge, W. R. (1959). The effects of increased filling pressure on rhythmicity and atrioventricular conduction in isolated hearts. *J. Physiol., Lond.* **149**, 193–208.

Kempton, R. T. (1969). Morphological features of functional significance in the gills of the spiny dogfish, *Squalus acanthias. Biol. Bull. mar. biol. Lab., Woods Hole,* **136**, 226–40.

Kirby, S. and Burnstock, G. (1969). Comparative pharmacological studies of isolated spiral strips of large arteries from lower vertebrates. *Comp. Biochem. Physiol.* **28**, 307–20.

Kirk, J. E. (1959). Mucopolysaccharides of arterial tissue. In *The Arterial Wall.* Ed. A. I. Lansing. London: Baillière, Tindall and Cox.

Kisch, B. (1948). Electrocardiographic investigation of the heart of fish. *Expl. Med. Surg.* **6**, 31–62.

Kisch, B. (1950). Reflex cardiac inhibition in the ganoid *Acipenser sturio. Am. J. Physiol.* **160**, 552–5.

Kishinouye, K. (1923). Contributions to the comparative study of the so-called Scombroid fishes. *Univs. Facul. Agric. Bull., Tokyo,* **8**, 283–475.

Klawe, W. L., Barrett, I. and Klawe, B. M. (1963). Haemoglobin content of the blood of six species of Scombroid fishes. *Nature, Lond.* **198**, 96.

Klontz, G. W. (1969). Immunogenesis in rainbow trout (*Salmo gairdneri*). (In preparation.)

Klontz, G. W., Yasutake, W. T. and Parisot, T. J. (1965). Virus diseases of the Salmonidae in western United States. III. Immunopathological aspects. *Ann. N.Y. Acad. Sci.* **126**, 531–542.

Klontz, G. W., Yasutake, W. T., Wales, J. E., Ashley, L. M. and Smith, C. (1969). The application of haematological technique to fishery research. (In preparation.)

Koehn, R. K. (1966). Serum haptoglobins in some North American catostomid fishes. *Comp. Biochem. Physiol.* **17**, 349–52.

Krogh, A. (1959). *The Anatomy and Physiology of the Capillaries.* New York: Hafner.

Kuhn, H. J., Moser, R. and Kuhn, W. (1962). Haarnadelgegenstrom als Grundlage zur Erzeugung hoher Gasdrücke in der Schwimmblase von Tiefseefischen. *Pflugers Archs. ges. physiol.* **275**, 231–7.

Kuhn, W., Ramel, A., Kuhn, H. J. and Marti, E. (1963). The filling mechanism of the swimbladder. *Experientia,* **19**, 497–511.

Kuriyama, H. A., Goto, M., Maéno, T., Abe, Y. and Ozaki, S. (1960). Comparative studies on transmembrane potentials and electrical characteristics of cardiac muscles. In *Electrical Activity of Single Cells,* pp. 243–60. Ed. Y. Katsuki. Tokyo: Izaku Shoin Ltd.

Kutty, M. N. (1968). Influence of ambient oxygen on the swimming performance of goldfish and rainbow trout. *Can. J. Zool.* **46**, 647–53.

Labat, R. (1966). *Electrocardiologie chez les poissons téléostéens: influence de quelques facteurs ecologique.* Toulose: Privat.

Laguesse, E. (1892). Bourrelets valvulaires artériels chez les poissons (*Labrus, Crenilabrus*). *C.R. Acad. Sci., Paris,* **44**, 211–13.

Lahlou, B., Henderson, I. W. and Sawyer, W. H. (1969). Renal adaptations by *Opsanus tau,* a euryhaline aglomerular teleost, to dilute media. *Am. J. Physiol.* **216**, 1266–72.

Lander, J. (1964). The shark circulation. B.Sc. Med. Dissertation. University of Sydney.

Lansing, A. I. (1959). Elastic tissue. In *The Arterial Wall*. Ed. A. I. Lansing. London: Baillière. Tindall and Cox.

Laurent, P. (1962). Contribution à l'étude morphologique et physiologique de l'innervation de coeur des téléostéens. *Archs. Anat. microsc. Morph. exp.* **51**, 337–458.

Laurent, P. (1967). Le pseudobranchie des Téléostéens: preuves electrophysiologique de ses fonctions, chémoréceptrice et baroréceptrice. *C.R. Acad. Sci. Paris*, **264**, 1879–82.

Laurent, P. and Dunel, S. (1964). L'innervation de la pseudobranchie chez la Tanche. *C.R. Acad. Sci.*, *Paris*, **258**, 6230.

Laurent, P. and Dunel, S. (1966). Recherches sur l'innervation de la pseudobranchie des téléostéenes. *Archs. Anat. microsc. Morph. exp.* **55**, 633–56.

Learoyd, B. M. (1963). Unpublished results.

Leiner, M. (1938). Die Augenkiemedrüse (Pseudobranchie) der Knochenfische. *Z. vergl. Physiol.* **26**, 416–66.

Lenfant, C. (1969). Personal communication.

Lenfant, C. and Johansen, K. (1966). Respiratory function in the elasmobranch *Squalus suckleyi* G. *Resp. Physiol.* **1**, 13–29.

Lenfant, C. and Johansen, K. (1969). Personal communication.

Lenfant, C., Johansen, K. and Grigg, G. C. (1966/67). Respiratory properties of the blood and pattern of gas exchange in the lung fish. *Neoceratodus forsteri* (Krefft). *Resp. Physiol.* **2**, 1–21.

Lenfant, C., Torrence, J., English, E., Finch, C. A., Reynafarge, C., Ramos, J. and Faura, J. (1968). Effect of altitude on oxygen binding by haemoglobin and on organic phosphate levels. *J. clin. Invest.* **47**, 2651–6.

Lenfant, C., Torrence, J. D., Woodson, R. D., Jacobs, P. and Finch, C. A. (1970). Role of organic phosphates in the adaptation of man to hypoxia. *Fed. Proc.* **29**, 115.

Lenhert, P. G., Lowe, W. E. and Carlson, F. D. (1956). The molecular weight of hemoglobin from *Petromyzon marinus*. *Biol. Bull. mar. biol. Lab.*, *Woods Hole*. **111**, 293–4.

Lepkovsky, S. (1929). The distribution of serum and plasma proteins in fish. *J. biol. Chem.* **85**, 667–73.

Lieb, J. R., Slane, G. M. and Wilber, C. G. (1953). Hematological studies on Alaska fish. *Trans. Am. microsc. Soc.* **72**, 37–47.

Lutz, B. R. and Wyman, L. C. (1932). Reflex cardiac inhibition of branchiovascular origin in the elasmobranch *Squalus acanthias*. *Biol. Bull. mar. biol. Lab.*, *Woods Hole*, **62**, 10–16.

Lyon, E. P. (1926). A study of the blood pressure, circulation and respiration of sharks. *J. gen. Physiol.* **8**, 279–87.

McKnight, I. M. (1966). A hematological study of the Mountain White fish *Prosopium williamsoni*. *J. Fish. Res. Bd Can.* **23**, 45–64.

Maetz, J. (1953). L'anhydrase carbonique dans deux Téléostéens voisins. Comparison des activites anhydrasiques chez *Perca* et *Serranus*. *C.R. Soc. Biol.*, *Paris*, **147**, 204–6.

Manwell, C. (1958a). On the evolution of hemoglobin. Respiratory properties of the hemoglobin of the California hagfish, *Polistotrema stouti*. *Biol. Bull. mar. biol. Lab.*, *Woods Hole*, **115**, 227–38.

Manwell, C. (1958b). Ontogeny of hemoglobin in the Skate, *Raja binoculata.*
Science, N.Y. **128**, 419–20.

Manwell, C. (1958c). A fetal-maternal shift in the ovoviviparous Spiny dogfish
Squalus suckleyi (Girard). *Physiol. Zoöl.* **31**, 93.

Manwell, C. (1964). Chemistry, genetics and function of invertebrate respiratory
pigments – configurational changes and allosteric effects. In *Oxygen in the
Animal Organism.* Ed. F. Dickens and E. Neil. New York: Pergamon Press.

Manwell, C. and Childers, W. (1963). The genetics of hemoglobin in hybrids.
I. A molecular basis for hybrid vigor. *Comp. Biochem. Physiol.* **10**, 103–20.

March, H. W., Ross, J. K. and Lower, R. R. (1962). Observations on the be-
haviour of the right ventricular outflow tract with reference to its develop-
mental origins. *Am. J. Med.* **32**, 835–45.

Marples, B. J. (1935–6). The blood vascular system of the elasmobranch fish
Squatina squatina (Linne). *Trans. R. Soc. Edinb.* **58**, 817–40.

Marshall, N. B. (1960). Swimbladder structure of deep-sea fishes in relation to
their systematics and biology. *'Discovery' Rep.* **31**, 1–122.

Martin, A. W. (1950). Some remarks on the blood volume of fish. In *Studies
Honouring Trevor Kincaid,* pp. 125–40. Ed. M. H. Hatch. Seattle: University
of Washington.

Marvin, D. E. and Heath, A. G. (1968). Cardiac and respiratory responses to
gradual hypoxia in three ecologically distinct species of freshwater fish.
Comp. Biochem. Physiol. **27**, 349–55.

Mayer, P. (1888). Über Eigentumlichkeiten in den Kreislauforganen der Selachiern.
Mitt. zool. Stn Neapel, **8**, 307–73.

Mitchell, J. H., Gilmore, J. P. and Sarnoff, S. J. (1962). The transport function
of the atrium. Factors influencing the relation between mean left atrial
pressure and left ventricular end diastolic pressure. *Am. J. Cardiol.* **9**,
237–47.

Moller, D. and Naevdal, G. (1965). Serum transferrins in some gadoid fishes.
Nature, Lond. **210**, 317–18.

Mott, J. C. (1950). Radiological observations on the cardiovascular system in
Anguilla anguilla. J. exp. Biol. **27**, 324–33.

Mott, J. C. (1951). Some factors affecting the blood circulation in the common
eel (*Anguilla anguilla*). *J. Physiol., Lond.* **114**, 387–98.

Mott, J. C. (1957). The cardiovascular system. In *The Physiology of Fishes.* Ed.
M. E. Brown, **1**, 81–108. New York: Academic Press.

Muir, B. S. and Hughes, G. M. (1969). Gill dimensions for three species of Tunny.
J. exp. Biol. **51**, 271–85.

Munshi, J. S. D. and Singh, B. N. (1968). On the microcirculatory system of the
gills of certain freshwater teleostean fishes. *J. Zool.* **154**, 365–76.

Nakano, T. and Tomlinson, N. (1967). Catecholamine and carbohydrate concentrations in rainbow trout (*Salmo gairdneri*) in relation to physical disturb-
ance. *J. Fish. Res. Bd Can.* **24**, 1701–15.

Nandi, J. (1962). The structure of the interrenal gland in teleost fishes. *Univ.
Calif. Publs Zool.* **65**, 129–212.

Nawar, G. (1955). On the anatomy of *Clarias lazera.* III. The vascular system.
J. Morph. **97**, 179–214.

Neuman, R. E. and Logan, M. A. (1950). The determination of collagen and
elastin in tissues. *J. biol. Chem.* **186**, 549–56.

Newstead, J. D. (1967). Fine structure of the respiratory lamellae of teleostean gills. *Z. Zellforsch, mikrosk. Anat.* **79**, 396–428.

Nicol, J. A. C. (1950). The autonomic nervous system of the Chimaeroid fish, *Hydrolagus colliei. Q. J. Micr. Sci.* **91**, 379–99.

Nicol, J. A. C. (1952). Autonomic nervous system in lower chordates. *Biol. Rev.* **27**, 1–49.

Nielsen, J. G. and Munk, O. (1964). A hadel fish (*Bassogigas profundissimus*) with a functional swim bladder. *Nature, Lond.* **204**, 594–5.

Oets, J. (1950). Electrocardiograms of fishes. *Physiologia comp. Oecol.* **2**, 181–6.

Ogden, E. (1945). Respiratory flow in *Mustelus. Am. J. Physiol.* **145**, 134–9.

Östlund, E. (1954). The distribution of catecholamines in lower animals and their effect on the heart. *Acta physiol. scand.* Suppl. **112**, 1–67.

Östlund, E. and Fänge, R. (1962). Vasodilation by adrenaline and noradrenaline and the effects of some other substances on perfused fish gills. *Comp. Biochem. Physiol.* **5**, 307–9.

Parry, G. and Holliday, F. G. T. (1960). An experimental analysis of the function of the pseudobranch in teleosts. *J. exp. Biol.* **37**, 344–54.

Parsons, C. W. (1930). The conus arteriosus in fishes. *Q. J. Micr. Sci.* **73**, 145–76.

Pathak, C. L. (1958). Effect of stretch on formation and conduction of electrical impulses in the isolated sinoauricular chamber of frog's heart. *Am. J. Physiol.* **192**, 111–13.

Perutz, M. F. (1964). The hemoglobin molecule. *Scient. Am.* **211**, November, 64–76.

Perutz, M. F., Muirhead, M., Cox, J. M. and Goaman, L. C. G. (1968). Three dimensional fourier synthesis of horse oxyhaemoglobin at a 2.8 Å resolution: the atomic model. *Nature, Lond.* **219**, 131–9.

Pflugfelder, O. (1952). Weitere volumetrische Untersuchungen über die Wirkung der Augenexstirpation und der Dunkelhaltung auf das Mesencephalon und die Pseudobranchien von Fischen. *Willhelm Roux Arch. Entw Mech.* **145**, 549–60.

Piiper, J. and Schumann, D. (1967). Efficiency of O_2 exchange in the gills of the dogfish, *Scyliorhinus stellaris. Resp. Physiol.* **2**, 135–48.

Prosser, C. L., Barr, L. M., Ping, R. D. and Lauer, C. Y. (1957). Acclimation of goldfish to low concentrations of oxygen. *Physiol. Zoöl.* **30**, 137–41.

Rahn, H. (1966). Aquatic gas exchange: theory. *Resp. Physiol.* **1**, 1–12.

Randall, D. J. (1966). The nervous control of cardiac activity in the tench (*Tinca tinca*) and the goldfish (*Carassius auratus*). *Physiol. Zoöl.* **39**, 185–92.

Randall, D. J. (1968). Functional morphology of the heart in fishes. *Am. Zoologist,* **8**, 179–89.

Randall, D. J., Holeton, G. F. and Stevens, E. D. (1967). The exchange of oxygen and carbon dioxide across the gills of rainbow trout. *J. exp. Biol.* **46**, 339–48.

Randall, D. J. and Jones, D. (1969). (In preparation.)

Randall, D. J., Jones, D. and Shelton, G. (1969). (In preparation.)

Randall, D. J. and Shelton, G. (1965). The effects of changes in environmental gas concentrations on the breathing and heart rate of a teleost fish. *Comp. Biochem. Physiol.* **9**, 229–39.

Randall, D. J. and Smith, J. C. (1967). The regulation of cardiac activity in fish in a hypoxic environment. *Physiol. Zoöl.* **40**, 104–13.

Randall, D. J. and Stevens, E. D. (1967). The role of adrenergic receptors in cardiovascular changes associated with exercise in salmon. *Comp. Biochem. Physiol.* **21**, 415–24.

Rapoport, S. and Guest, G. M. (1941). Distribution of acid soluble phosphorus in the blood cells of various vertebrates. *J. biol. Chem.* **138**, 269–82.

Read, J. B. and Burnstock, G. (1968). Comparative histo-chemical studies of adrenergic nerves in the enteric plexuses of vertebrate large intestine. *Comp. Biochem. Physiol.* **27**, 505–17.

Reznikoff, P. and Reznikoff, G. (1934). Haematological studies in dogfish (*Mustelus canis*). *Biol. Bull. mar. biol. Lab.*, Woods Hole, **66**, 115–23.

Riggs, A. (1965). Functional properties of hemoglobin. *Physiol. Rev.* **45**, 619–73.

Roach, M. R. and Burton, A. C. (1957). The reason for the shape of the distensibility curves of arteries. *Can. J. Biochem. Physiol.* **35**, 681–90.

Roberts, B. L. (1969). Spontaneous rhythms in the motoneurons of spinal dogfish (*Scyliorhinus canicula*). *J. mar. biol. Ass. U.K.* **49**, 33–49.

Robertson, O. H., Krupp, M. A., Thompson, N., Thomas, S. F. and Hane, S. (1966). Blood pressure and heart weight in immature and spawning pacific salmon. *Am. J. Physiol.* **210**, 957–64.

Rodionov, I. M. (1959). Reflex control of the heart in fish. II. Reflex effect on the heart and intestinal blood vessels in stimulation of the baroceptors of the gill blood vessels. *Bull. exp. Biol. Med. U.S.S.R.* **47**, 653–6.

Root, R. W. (1931). The respiratory function of the blood of marine fishes. *Biol. Bull. mar. biol. Lab.*, Woods Hole. **61**, 427–56.

Root, R. W. and Irving, L. (1941). The equilibrium between hemoglobin and oxygen in whole and hemolyzed blood of the tautog, with a theory of the Haldane effect. *Biol. Bull. mar. biol. Lab.*, Woods Hole, **81**, 307–23.

Ruud, J. T. (1954). Vertebrates without erythrocytes and blood pigment. *Nature, Lond.* **173**, 848–50.

Rybak, B. and Cortot, H. (1956). La valvule sino-auriculaire, isolée du coeur de *Scyllium canicula*, preparation de choise pour l'étude de l'automatism myocardique. *C.R. Soc. Biol., Paris*, **150**, 2216–18.

Sano, T. (1960). Change in blood constituents with growth of rainbow trout. *J. Tokyo Univ. Fish.* **46**, 77–87.

Satchell, G. H. (1960). The reflex coordination of the heart beat with respiration in the dogfish. *J. exp. Biol.* **37**, 719–31.

Satchell, G. H. (1961). The response of the dogfish to anoxia. *J. exp. Biol.* **38**, 531–43.

Satchell, G. H. (1962). Intrinsic vasomotion in the dogfish gill. *J. exp. Biol.* **39**, 503–12.

Satchell, G. H. (1965). Blood flow through the caudal vein of elasmobranch fish. *Aust. J. Sc.* **27**, 240–1.

Satchell, G. H. (1968). The genesis of certain cardiac arrhythmias in fish. *J. exp. Biol.* **49**, 129–41.

Satchell, G. H. (1969). Unpublished observations.

Satchell, G. H. (1970). A functional appraisal of the fish heart. *Fed. Proc.* **29**, 1120–3.

Satchell, G. H. and Jones, M. P. (1967). The function of the conus arteriosus in the Port Jackson shark, *Heterodontus portusjacksoni*. *J. exp. Biol.* **46**, 373–82.

Satchell, G. H., Hanson, D. and Johansen, K. (1970). Differential blood flow through the afferent branchial arteries of the skate, *Raja rhina*. (In the press.)

Saunders, D. C. (1966a). Elasmobranch blood cells. *Copeia*, **2**, 348–51.

Saunders, D. C. (1966b). Differential blood cell counts of 121 species of marine fishes of Puerto Rico. *Trans. Am. microsc. Soc.* **85**, 427–49.

Saunders, D. C. (1967). Neutrophils and arneth counts from some Red Sea fishes. *Copeia*, **3**, 681–3.

Saunders, D. C. (1968). Differential blood cell counts of 50 species of fishes from the red sea. *Copeia*, **3**, 491–8.

Saunders, R. L. (1962). The irrigation of the gills in fishes. II. Efficiency of oxygen uptake in relation to respiratory flow, activity and concentrations of oxygen and carbon dioxide. *Can. J. Zool.* **40**, 817–62.

Saxena, D. B. and Bakhshi, P. L. (1965). Cardio-vascular system of some fishes of the torrential streams in India. Part 1. Heart of *Orienus plagiostomus plagiostomus* and *Botia birdi*. *Jap. J. Ichthyol.* **12**, 70–81.

Schoenlein, K. (1895). Beobachtungen über Blutkreislauf und Respiration bei einigen Fischen. *Z. biol.* **32**, 511–47.

Scholander, P. F. (1954). Secretion of gases against high pressures in the swim bladder of deep sea fishes. II. The rete mirabile. *Biol. Bull. mar. biol. Lab., Woods Hole*, **107**, 260–77.

Scholander, P. F., and Van Dam, L. (1954). Secretion of gases against high pressures in the swim bladder of deep sea fishes. I. Oxygen dissociation in blood. *Biol. Bull. mar. biol. Lab., Woods Hole*, **107**, 247–59.

Seyama, I. and Irisawa, H. (1967). The effect of high sodium concentration on the action potential of the skate heart. *J. gen. Physiol.* **50**, 505–17.

Shelton, G. and Randall, D. J. (1962). The relationship between heart beat and respiration in teleost fish. *Comp. Biochem. Physiol.* **7**, 237–50.

Shepherd, D. M., West, G. B. and Erspamer, V. (1953). Chromaffin bodies of various species of dogfish. *Nature, Lond.* **112**, 509.

Shepherd, M. P. (1955). Resistance and tolerance of young speckled trout (*Salvelinus fontinalis*) to oxygen lack with special reference to low oxygen acclimation. *J. Fish. Res. Bd Can.* **12**, 387–433.

Sick, K., Frydenberg, O. and Nielsen, J. T. (1963). Haemoglobin patterns of plaice, flounder and their natural and artificial hybrids. *Nature, Lond.* **198**, 411–12.

Smith, H. W. (1929). The composition of the body fluids of elasmobranchs. *J. biol. Chem.* **81**, 407–19.

Smith, L. S. (1966). Blood volumes of three salmonids. *J. Fish. Res. Bd Can.* **23**, 1439–46.

Smith, L. S. and Bell, G. R. (1964). A technique for prolonged blood sampling in free swimming salmon. *J. Fish. Res. Bd Can.* **21**, 711–17.

Smith, L. S., Brett, J. R. and Davis, J. C. (1967). Cardiovascular dynamics in swimming adult sockeye salmon. *J. Fish. Res. Bd Can.* **24**, 1775–90.

Smith, L., Lewis, W. M. and Kaplan, H. H. (1952). Comparative morphological and physiological study of fish blood. *Progve Fish Cult.* **14**, 169–72.

Smith, W. C. (1918). On the process of disappearance of the conus arteriosus in teleosts. *Anat. Rec.* **15**, 65–71.

Steen, J. B. (1963). The physiology of the swimbladder in the eel *Anguilla vulgaris*, III. The mechanism of gas secretion. *Acta physiol. scand.* **59**, 221–41.

Steen, J. B. and Berg, T. (1966). The gills of two species of haemoglobin free fishes compared to those of other teleosts – with a note on severe anaemia in an eel. *Comp. Biochem. Physiol.* **18**, 517–26.

Steen, J. B. and Kruysse, A. (1964). The respiratory function of teleostean gills. *Comp. Biochem. Physiol.* **21**, 127–42.

Stevens, E. D. (1968). The effect of exercise on the distribution of blood to various organs in rainbow trout. *Comp. Biochem. Physiol.* **25**, 615–25.

Stevens, E. D. and Black, E. C. (1966). The effect of intermittent exercise on carbohydrate metabolism in rainbow trout, *Salmo gairdneri*. *J. Fish Res. Bd Can.* **23**, 471–85.

Stevens, E. D. and Randall, D. J. (1967a). Changes in blood pressure, heart rate and breathing rate during moderate swimming activity in rainbow trout. *J. exp. Biol.* **46**, 307–15.

Stevens, E. D. and Randall, D. J. (1967b). Changes in gas concentrations in blood and water during moderate swimming activity in rainbow trout. *J. exp. Biol.* **46**, 329–37.

Stohr, P. (1876). Ueber den Klappenapparat im Conus arteriosus der Selachier und Ganoiden. *Morph. Jb.* **2**, 197–228.

Sudak, F. N. (1965a). Intrapericardial and intracardiac pressures and the events of the cardiac cycle in *Mustelus canis* (Mitchill). *Comp. Biochem. Physiol.* **14**, 689–705.

Sudak, F. N. (1965b). Some factors contributing to the development of sub-atmospheric pressure in the heart chambers and pericardial cavity of *Mustelus canis* (Mitchill). *Comp. Biochem. Physiol.* **15**, 199–215.

Sudak, F. N. and Wilber, C. G. (1960). Cardiovascular responses to hemorrhage in the dog fish. *Biol Bull. mar. biol. Lab., Woods Hole*, **119**, 342.

Sulya, L. L., Box, B. E. and Gunter, G. (1960). Distribution of some blood constituents in fishes from the Gulf of Mexico. *Am. J. Physiol.* **199**, 1177–80.

Sulya, L. L., Box, B. E. and Gunter, G. (1961). Plasma proteins in the blood of fishes from the Gulf of Mexico. *Am. J. Physiol.* **200**, 152–4.

Sutterlin, A. M. (1969). Effects of exercise on cardiac and ventilation frequency in three species of freshwater teleost. *Physiol. Zoöl.* **42**, 36–52.

Taylor, M. G. (1964). Wave travel in arteries and the design of the cardiovascular system. In *Pulsatile Blood Flow.* Ed. E. O. Attinger. New York: McGraw-Hill.

Taylor, W., Houston, A. H. and Horgan, J. D. (1968). Development of a computer model simulating some aspects of the cardiovascular respiratory dynamics of the salmonid fish. *J. exp. Biol.* **49**, 477–93.

Tebecis, A. K. (1967). A study of electrograms recorded from the conus arteriosus of an elasmobranch heart. *Aust. J. biol. Sci.* **20**, 843–6.

Thorson, T. B. (1958). Measurements of the fluid compartments of four species of marine chondrichthyes. *Physiol. Zoöl.* **31**, 16–23.

Thorson, T. B. (1959). Partitioning of body water in Sea Lamprey. *Science, N.Y.* **130**, 99–100.

Thorson, T. B. (1961). The partitioning of body water in osteichthyes; phylogenetic and ecological implications in aquatic vertebrates. *Biol. Bull. mar. biol. Lab., Woods Hole*, **120**, 238–54.

Tyler, J. C. (1960). Erythrocyte counts and haemoglobin determinations for two antarctic nototheniid fishes. *Stanford ichthyol. Bull.* **7**, 199–201.

Wagenvoort, C. A. (1952). *De functie van de arteriële ringen: een morphologisch en experimenteel onderzoek.* Utrecht: N. V. Drukker I.J.P. den Boer.

Wald, G. and Riggs, A. (1951). The hemoglobin of the sea lamprey, *Petromyzon marinus. J. gen. Physiol.* **35,** 45–53.

West, G. B. (1955). The comparative pharmacology of the suprarenal medulla. *Q. Rev. Biol.* **30,** 116–37.

White, F. W. (1969). Unpublished observations.

White, F. W. and Satchell, G. H. (1969). Unpublished observations.

Wiedmann, S. (1955). Effects of calcium ions and local anaesthetics on electrical properties of Purkinje fibres. *J. Physiol., Lond.* **129,** 568–82.

Willmer, E. N. (1934). Some observations of the respiration of certain tropical freshwater fishes. *J. exp. Biol.* **11,** 283–306.

Wintrobe, M. M. (1934). Variations in the size and haemoglobin content of erythrocytes in the blood of various vertebrates. *Folia haemat., Lpz.* **51,** 32–49.

Wittenberg, J. B. and Wittenberg, B. A. (1962). Active secretion of oxygen into the eye of fish. *Nature, Lond.* **194,** 106–7.

Young, J. Z. (1931). On the autonomic nervous system of the teleostean fish. *Uranoscopus scaber. Q. J. Micr. Sci.* **74,** 491–535.

Young, J. Z. (1933). The autonomic nervous system of Selachians. *Q. J. Micr. Sci.* **75,** 571–624.

Zanjani, E. D., Yu, M. L., Perlmutter, A. and Gordon, A. D. (1969). Humoral factors influencing erythropoiesis in the fish, blue gourami, *Trichogaster trichopterus. Blood,* **33,** 573–81.

Zwaardemaker, H. and Noyens, A. K. M. (1910). Das Elektrogramm des isoliert pulsierenden Aal ventrikels. *Onderz. Physiol. Lab. Rijksuniv., Utrecht.* **10,** 155–91.

INDEX